The last great plant hunt

The last great plant hunt

The story of Kew's Millennium Seed Bank

Carolyn Fry, Sue Seddon and Gail Vines

Kew Publishing

Royal Botanic Gardens, Kew

Great care has been taken to maintain the accuracy of the information contained in this
work. However, neither the publishers nor the authors can be held responsible for any
consequences arising from use of the information contained herein.

First published in 2011 by
Royal Botanic Gardens, Kew
Richmond, Surrey, TW9 3AB, UK

www.kew.org

ISBN 978 1 84246 432 8

British Library Cataloguing in Publication Data
A catalogue record for this book is available from the British Library.

Main cover photograph: Jeff Eden
Other cover photographs: Andrew McRobb

Editor: Michelle Payne
Design, typesetting and page layout: Jeff Eden
Cover design: Jeff Eden

Printed and bound by Firmengruppe APPL, aprinta druck,
Wemding, Germany

For information or to purchase all Kew titles please visit
www.kewbooks.com or email publishing@kew.org

Kew's mission is to inspire and deliver science-based plant conservation worldwide,
enhancing the quality of life.

Kew receives half of its running costs from Government through the Department for
Environment, Food and Rural Affairs (Defra). All other funding needed to support
Kew's vital work comes from members, foundations, donors and commercial activities
including book sales.

Frontispiece:
Ulises Guzman collecting seeds in Mexico. Plants from here and many other countries
are now safeguarded thanks to the efforts of the Millennium Seed Bank Partnership.

The paper used in this book contains material sourced from responsibly managed and
sustainable commercial forests, certified in accordance with the FSC (Foresty Stewardship
Council), and manfactured under strict environmental systems, the international
140001standard, EMAS (Eco-Management and Audit Scheme) and the IPPC (Integrated
Pollution Prevention and Control) regulation.

Contents

Foreword

by HRH Prince of Wales 6

1 What is the last great plant hunt?

A quest to save biodiversity before it is too late 10

Insight: Nature's life-giving works of art 14

How seeds underpin human evolution 16

Treasure Hunt: Tsodilo daisy 20

Banking wild seeds gives an insurance policy for plants 22

Job Profile: Wolfgang Stuppy – Seed Morphologist 26

2 Conserving wild plants on a global scale

How Kew Gardens became a powerhouse of plant science 30

Reaching out to save seeds and halt biodiversity loss 34

Insight: A short history of banking seeds 38

Insight: The history of seed banking at Kew 40

Two types of seeds pose a quandary for conservation 42

The three 'E's provide a priority for saving 44

Treasure Hunt: Starfruit 46

Job Profile: Paul Smith – Head of the Seed Conservation Unit 48

3 In search of the world's seeds

Finding the best places to collect seeds 52

Insight: International partnerships 56

The band of botanists who forge partnerships 58

Job Profile: Tim Pearce – International Coordinator 60

Job Profile: Michael Way – International Coordinator 62

Job Profile: Michiel van Slageren – International Coordinator 64

Job Profile: Moctar Sacande – International Coordinator 66

Job Profile: Clare Trivedi – International Coordinator 68

Treasure Hunt: Syrian bear's breeches 70

Local botanists build a global seed bank 72

Job Profile: Dan Duval – Seed Collector 74

How botanists gather seeds from the world's precious flora 76

Tracking down a lost plant in China's limestone country 78

Bringing the world's rare orchids back from the brink of extinction 80

Desert mission safeguards rarities 82

Scouring Chile's hills for endangered botanical bounty 86

4 The science of saving seeds for posterity

Inside the most biodiverse building on planet Earth 92

Job Profile: Keith Manger – Laboratory and Building Manager 96

Insight: The journey of a seed through the Seed Bank 100

Preparing seeds for storage in the subterranean seed vault 102

Job Profile: John Adams – Technology Expert 110

Cutting-edge seed research 114

Unravelling the mystery of how long seeds live 116

Treasure Hunt: Botanic booty 120

…springs to life 122

The quest for the germination predictor 124

Job Profile: Robin Probert – Head of Seed Conservation and Technology 126

Understanding how seeds rely on seasonal shifts 128

To dry or not to dry? Understanding recalcitrant seeds 130

Job Profile: Hugh Pritchard – Head of Research 132

Searching for the elixir of plant life 134

5 Sharing knowledge and putting seeds to work

Exchanging seed secrets with global partners 140

Job Profile: Vanessa Sutcliffe – Technology Specialist 142

Job Profile: Kate Gold – Training Manager 144

Sharing seed knowledge boosts local livelihoods 146

Insight: Where are they now – former students around the world 148

SID shares data with others 150

Growing useful plants in villages around the globe 152

Useful Plants Project: safeguarding plants for local communities 154

Restoring habitats in South Africa 158

Restoring habitats in the Middle East 160

Restoring habitats in Australia and the USA 162

Interviews around the world: Jie Cai, Chin 164

Interviews around the world: Masego Kruger, Botswana 166

Treasure Hunt: Willowmore cedar 168

Propagating threatened cacti and bulbs in Chile 170

Job Profile: Rosemary Newton – Seed Germination Specialist 172

6 Breathing life into degraded ecosystems

Saved seeds spring to life and restore damaged habitats 176

Helping partners use 'difficult' seeds 180

Founding heroes of the Millennium Seed Bank Project 184

Saving the endangered triangular club rush 186

Find out more 188

Index 190

Acknowledgements/Photography credits 192

As Patron of The Foundation and Friends of The Royal Botanic Gardens, Kew, it gives me great pleasure to contribute this foreword to 'The Last Great Plant Hunt'.

Seeds are one of Nature's marvels; they are cunning and highly evolved to meet all manner of environmental challenge. They are beautifully exquisite at microscopic level. They have a long, interwoven association with the evolution, history and culture of our own species and they still provide us with our most basic source of food. Increasingly, seeds play a vital role, not only in ensuring the survival of the plants that expend enormous energies to produce them, but in the survival of our planet and humankind itself.

When I opened Kew's Millennium Seed Bank, in 2000, I described the project as the 'Bank of England of the botanical world' and, tragically, the need for it has continued to grow. Today it is estimated that more than 20% of all plants – that is over 76,000 species – are threatened with extinction as their habitats shrink. Nowhere is this clearer than in the rainforests of the world, which are being destroyed or degraded by 50 million acres (over 20 million hectares) per year.

Plants are absolutely fundamental to life; they provide us with the air we breathe, help to supply our water, the houses we live in, the food we eat and the medicines that heal us. This was made very clear to me recently by Dr. Eric Chivian, the Nobel Peace Prize-winner from Harvard Medical School and one of the leading scientists looking at biodiversity. He has demonstrated conclusively that there is a direct relationship between the health of humans and the levels of biodiversity in the world – whether it is the destruction of the world's rainforests, which provide the vital rainfall on which global agriculture depends, or the loss of natural organisms that has a direct effect on the spread of infectious diseases.

Most people are familiar with the Royal Botanic Gardens, Kew, as a World Heritage Site, with stunning glasshouses, other architecturally important buildings and aesthetic landscapes. But Kew also works with hundreds of partners around the world, identifying and studying plants in their habitats and assessing their complex relationship to human life and well-being. This knowledge has never been more important as the world population grows exponentially and as we face the consequences of climate change.

'The Last Great Plant Hunt' captures the essence of Kew's work, revealing in detail, for the first time, the story of the Millennium Seed Bank Project and Partnership. It tells in everyday language, the fascinating science and roles of the plant scientists and seed collectors in 50 partner countries. They have already banked seeds from ten per cent of the world's plants and are now striving to collect an unprecedented twenty-five per cent.

As Patron of the Kew Foundation, I am delighted to be contributing the foreword to this book and hope that you, as the reader, will be inspired enough to support this vital work.

Chapter 1

What is the last great plant hunt?

A cornucopia of fruits and
seeds received by the
Millennium Seed Bank
between 2000 and 2010.

A personal view from Paul Smith, Kew's Head of Seed Conservation

A quest to save biodiversity before it is too late

When I was a small boy, I was an avid collector of stamps. It's hard to say precisely what my motivation was, but I remember the thrill of buying a packet of assorted stamps and not knowing where they were going to be from. I would then spend happy hours finding the countries in my atlas, and sorting the stamps according to their country and age. The stamps were tangible small pieces of history and geography. This, combined with the acquisitive instinct that is in most of us, made stamp collecting the perfect hobby. Although my enthusiasm had waned by the time I was about ten, I do remember one particular set of stamps that the Royal Mail released called the Explorer series, which featured portraits of men such as Burton, Ross, Livingstone and Shackleton. These fully-bearded men of adventure had travelled to the ends of the Earth in the name of science and commerce. On the way they fought off lions, traversed freezing, wind-torn oceans in small boats, and disguised themselves as natives to avoid having their throats cut. The sense of adventure and the delicious range of possibilities instilled by the life stories of these explorers have stayed with me for much longer than the urge to collect stamps, and it is this spirit that continues in *The Last Great Plant Hunt*.

As a botanist and seed collector I have had the good fortune to study and collect plants in some of the world's most remote, beautiful and exciting places: the Mountains of the Moon (Uganda); the Kalahari desert (Botswana); the Great Lofty Mountains (Australia); the alpine meadows of Yunnan (China); the tall grass prairie (USA); and the islands of the Mediterranean (Crete). I have had my share of adventures, including being chased by elephants, attacked by men with machetes and falling ill with various tropical diseases. But here the comparison ends. Unlike Livingstone or Burton, modern plant hunters rarely meet new cultures and peoples for the first time; we work with them as partners instead. We don't take readings with sextants or make the maps as we go; we use Ordnance Survey charts and geographical positioning systems accurate to the metre. And we aren't collecting these plants for the first time; we are following in the footsteps of our forebears and the descriptions they have left us. Most importantly, perhaps, the work that we do is truly urgent because the world has changed drastically since the days of the great explorers.

The dominance of man on this planet was already becoming clear at the time of Shackleton's ordeal on South Georgia. The first national parks and nature reserves were being set aside, and the first laws were being passed to afford protection to wildlife. One hundred years on, it is clear that approach has been only partially successful. An estimated twelve per cent of the world's land surface area is formally protected to some degree, but the rest has been burnt, cut down, cleared and converted to feed and house a burgeoning human population. In Livingstone's day, there were less than two billion people on the planet. Today there are nearly seven billion, and by 2050 there will be over nine billion. The inevitable result is that there is less room for our fellow inhabitants. The extinction of species like the dodo, elephant bird, Tasmanian tiger and pink pigeon have made little impact, but more charismatic species, such as the panda, gorilla and tiger currently stand on the brink. If we lose any of these species through our own carelessness, we will undoubtedly mourn their passing, although the impact on humanity will be small.

With plants, the opposite is true. To the majority of people, plants are not charismatic: they aren't warm and cuddly, and they don't have big eyes that ask questions. And yet these countless, non-descript plants have crucial roles in maintaining life on this planet. They sit at the base of the food chain, providing food all the way up to us at the top. They provide services such as climate regulation and flood defence. They contribute to soil formation and nutrient cycling, and they provide us with shelter, medicines and fuel. Despite this, the Millennium Ecosystem Assessment estimates that between one quarter and one third of the total number of known plant species – that is 60,000 to 100,000 species – are threatened with extinction. The main threats are land-use change and over-exploitation, with climate change expected to exacerbate the situation.

The MSB and the MSBP

The Millennium Seed Bank (MSB) is the actual building and seed vault located at Wakehurst Place, Sussex. The Millennium Seed Bank Project (MSBP) refers to the first 10-year programme, which completed in 2010. The next phase – the Millennium Seed Bank Partnership (MSBP) – started when the Project finished.

Paul Smith manages Kew's Seed Conservation Department and the MSBP programme.

The destruction of the Amazon rainforest is causing species to be lost to extinction before we have even had a chance to name them and investigate their potential uses.

The threats facing the world's flora are a matter of grave concern for a number of reasons. Firstly, these plants may well be useful to us in unknown ways. The American naturalist, Aldo Leopold, wrote more than 60 years ago:

> If the biota, in the course of aeons, has built something we like but do not understand, then who but a fool would discard seemingly useless parts? To keep every cog and wheel is the first precaution of intelligent tinkering.

Since Leopold penned those words, the scientific discipline of ecology has demonstrated time and again that all productive systems are built on a web of interrelatedness. This is manifest in the simple relationships between plants, pollinators, pests and predators in our agricultural systems but it is true of all ecosystems, including the planet as a whole. We humans are not exempt from this. We are at the centre of this planet's ecology, and are becoming more and more dominant. A seemingly irrelevant plant may be essential to the life cycle of a pollinator. It may be the host of a useful fungus or it may be home to an insect or bird that keeps a crop pest in check. We condemn plants to extinction at our peril.

A second reason why we should care is because ecology has also taught us that resilience is found in diversity. The farmer who plants just one crop is far more susceptible to the vagaries of climate or disease than the farmer who plants a range of crops with a range of requirements and susceptibilities. The problem is that as a species we have forgotten this. Increasingly, we rely on simpler systems and a rapidly dwindling range of plant diversity. Eighty per cent of our calorie intake comes from just twelve plant species: eight grains and four tubers. And this is despite the fact that at least 30,000 species of plants are edible. Foresters primarily use around 100 tree species which are well-known throughout the world, such as species of *Pinus* and *Eucalyptus*, despite the fact that there are 60,000 species of tree we could be using. In Western medicine, we have only screened 20 per cent of the world's plant species for pharmaceutical activity even though 75 per cent of the world's population relies on wild plants for their primary health care. As the world grapples with the big environmental challenges of our day – food security, water availability, less land, climate change, deforestation, overpopulation, energy – we have to ask ourselves whether we can continue to rely on such a tiny fraction of the world's plant diversity for all of our future needs. Logic suggests that we cannot. We will need new food crops that use less water or that are resilient to

climate change. We will need to reforest catchment areas with more complex mixtures of trees that are not susceptible to pests and diseases. And we will need to develop first generation biofuels that do not displace food crops.

Finally, we should care about plant species becoming extinct because it is perfectly possible for us to save them. With the range of techniques available to us, there is no technological reason why any plant species should become extinct. Where possible, we should be protecting and managing plant populations *in situ* (in the wild). Where we can't do this, we should be keeping them in seed banks and in gardens. It is our responsibility – the responsibility of this generation – to give our children every opportunity we can, and that means safeguarding and passing on our biological inheritance intact. We will be judged not just by what we build, but by what we leave behind.

The Millennium Seed Bank Partnership epitomises this philosophy. Over the past 10 years, this partnership, comprising more than 120 plant science institutions in 50 countries, has successfully collected and secured in safe storage seeds from one in ten of the world's plant species. *The Last Great Plant Hunt* is the story of the Millennium Seed Bank Partnership, a global network that continues to go from strength to strength. Our next milestone is to have 25 per cent of the world's plant species stored as seed by 2020. As you will see from the narrative in this book, we don't just bank the seeds for posterity. We actively work on every single collection, finding out how useful it is and how we can grow it to enable human innovation, adaptation and resilience. We also send out our seeds to every corner of the planet to underpin vital research and development in agriculture, horticulture, forestry and restoration ecology.

I hope you will agree that, adventure notwithstanding, this is a long way from stamp collecting.

Nature's life-giving works of art

Seeds come in all shapes and sizes. Some tropical rainforest orchids produce tiny seeds, weighing only tiny fractions of a gram. The world's largest seed is that of the coco de mer palm, which can weigh up to 27 kg.

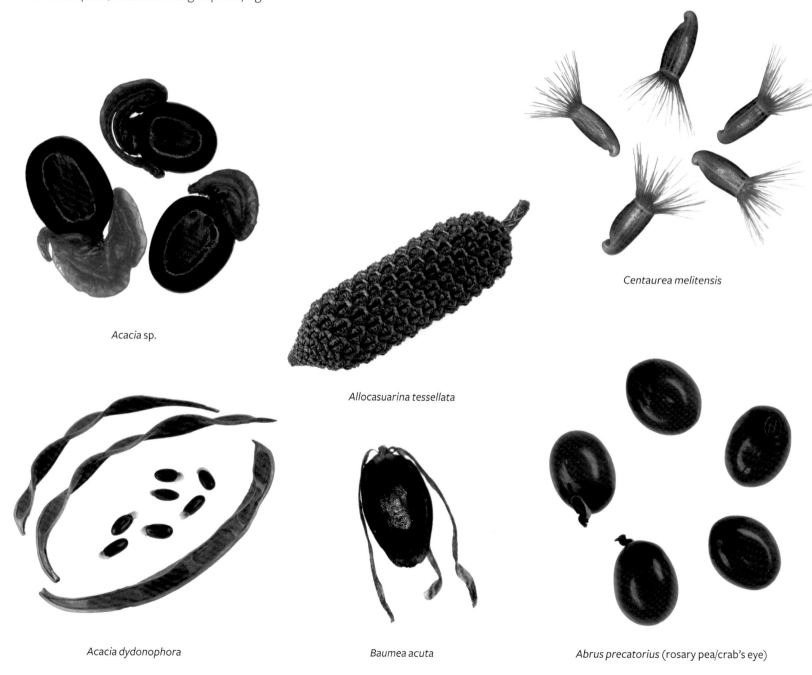

Acacia sp.

Allocasuarina tessellata

Centaurea melitensis

Acacia dydonophora

Baumea acuta

Abrus precatorius (rosary pea/crab's eye)

Strelitzia victoria-reginae
(Bird of paradise flower)

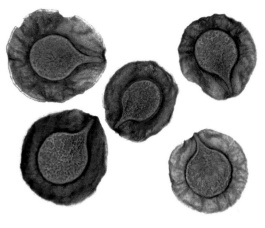

Velleia rosea

Aulax palasia

Acacia rosseii (in their pods)

Coffea arabica (coffee bean)

Stipagrostis giessi

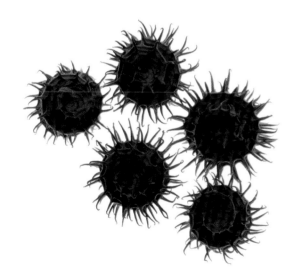

Medicago sp.

Lodoicea maldivica (Seychelles nut/
double coconut/coco de mer palm)

How seeds underpin human evolution

Our reliance on fruits and nuts has a long, much-debated, history. The ancestors of humans must have relied on fruits and seeds for their nutrition, just as chimpanzees do now. What may have separated the men from the monkeys, however, is the use of fire to make those nuts more palatable and digestible. If this was the case, it would have enabled our ape-like ancestors to consume food more quickly and digest it with less energy, with the effect of producing bigger bodies and brains. A 50 per cent larger brain and smaller teeth is what separates *Homo erectus*, our ancestor that appeared between 1.6 and 1.9 million years ago, from its predecessor *H. habilis*. Although direct evidence for the use of fire emerges later, some experts believe cooking foods caused those adaptations. So our dependence on seeds, and our ability to cook them, may just be what helped us evolve into the most intelligent species on Earth today.

The earliest evidence of humans using fire to cook seeds comes from an archaeological site in Israel, dated at 790,000 years ago. Archaeologists sifted through 23,454 seeds and fragments of fruit, plus 50,582 pieces of wood, looking for burned specimens at the site at Gesher Benot Ya'aqov in Northern Israel. They concluded that the small number of charred food and wood fragments they found was evidence that fire at the site had been controlled, rather than simply wildfire, and they interpreted clusters of burned flints as hearths. The researchers also found seven species of edible hard nuts, including almonds (*Amygdalus communis*) and acorns (*Overcus* sp.), along with pitted hammers and anvils that would have been needed to crack them. These findings provide early evidence of hommid diets and the technologies used to process foods. The tools uncovered at the site are similar to those used by chimpanzees and contemporary hunter-gatherers.

Much of the evidence of what people ate before the onset of farming is indirect. It is based on archaeologists' studies of teeth, where people lived and the known vegetation at those locations. One site in Syria, however, provides not only good evidence of the hunter-gatherer diet, but also how this changed once agriculture was adopted. The site, at Abu Hureyra in Syria, was inhabited between 11,500 to 7,000 years ago. Archaeological evidence shows that at 11,500 years ago, a hunter-gatherer community occupied the site in round houses. Because they were living in a biologically rich area they were able to live sedentary lives rather than be nomadic. They would have gathered seeds and fruits from locally available wild plants all year round. Archaeologists identified the charred remains of 142 species of plants, 118 of which comprise the seeds and hard fruits eaten by the hunter-gatherers. Among the species found were wild einkorn wheat and two types of wild rye.

Rubbing stones, querns, grinding dishes, mortars and pestles found at Abu Hureyra show that hard seeds were processed before being eaten. Pounding and grinding food to small particles ensures that food is properly digested and useful nutrients are absorbed. Not having to gather so much food from the surrounding areas would ultimately save on the amount of energy required by the hunter-gatherers. Experiments suggest that the uncovered querns and grinding stones would have enabled bulk processing of hard seeds, while pestles and mortars would have helped crack open nuts, such as acorns, hazel nuts and almonds. The community would have supplemented

CLOCKWISE FROM TOP LEFT:
The Millennium Seed Bank Partnership helps people around the world use plants in a beneficial and sustainable way. A lady grinds maize flour in a village near the Mwaluganje Reserve, Kenya; sorghum harvest in Burkina Faso, in 1974 (© Ray Witlin/ World Bank); close-up of winter wheat; *Bixa orellaria* seeds in Mali.

its diverse plant-based diet with small amounts of meat from animals such as gazelle. The spectrum of identified plants probably represents a proportion of those actually eaten, as softer leaf or root material would not have been preserved.

Evidence from seeds shows that the use of wild plant foods at Abu Hureyra declined rapidly around 11,050 years ago; at this point the appearance of a weed flora typical of arid-zone cultivation may be evidence for farming of wild plants. Some experts believe that environmental change to cooler temperatures in the Middle East may have caused wild plants to decline, prompting a move to cultivation. By 9,860 years ago, in the Neolithic age, a suite of domesticated crops including einkorn wheat, emmer wheat and lentils were being cultivated. Some wild-gathered feather-grass grain plus seeds of club-rush and Euphrates knot-grass continued to be used alongside the cultivated crops but their use slowly declined. By 8,500 years ago, evidence shows that the community was largely dependent on about eight species of domesticated plants for vegetable-based energy foods. This shift from wild-gathered to cultivated plants represents a significant narrowing of dietary diversity, but a significant increase in productivity.

Between 10,000 and 3,500 years ago, several human populations independently made the transition from a hunter-gatherer lifestyle to one based on agriculture. Changes to the climate are likely to have contributed to this switch in at least some locations. The centres in which agriculture originated, and their primary staple foods, are: south-west Asia around the Fertile Crescent (barley and wheat); China (rice and millet); Papua New Guinea (root and tree crops); sub-Saharan Africa (sorghum and pearl millet); Mesoamerica (maize and beans); eastern North America (several seed plants) and South America (quinoa and beans). These new agricultural economies enabled people to live at much higher densities and fuelled the creation of ever larger and more complex human societies. Villages soon turned into towns, and towns into cities. In time, city authorities began controlling the growing agricultural landscapes. As our ancestors became increasingly successful at organising agricultural production and the populations it fed, empires emerged.

The ability of nations to control large parts of the world, coupled with our ability to survive on a limited number of filling, energy-giving staple foods, gave rise to plantation-style agriculture from the 17th century. Oil palm is just one example of a crop grown in this way for its fruit and seeds, which provide oil for food and soaps. The better known of two species of oil palm, *Elaeis guineensis*, originated in Africa. From 1900 it was grown on European-run plantations in Central Africa and South-East Asia. The invention of hydrogenation provided new uses, such as in margarine, and fuelled a growing global industry. Between 1962 and 1982 world exports of palm oil rose from about 500,000 to 2,400,000 tonnes per annum, with Malaysia accounting for 85 per cent of those exports in 1982. In 2008, global production of palm oil and palm kernel oil stood at 48 million tonnes. As nations seek alternatives to fossil fuels, palm oil is increasingly being used to provide biodiesel.

Scientific advances in selective breeding and genetic modification (GM) have accompanied the globalisation of agriculture. These have tended to focus on creating crop varieties that enhance agriculturalists' abilities to grow crops efficiently on a vast scale; GM technology is often linked to dependence on other products from the same manufacturers. Varieties of sugar beet and oilseed rape, for example, have been genetically modified to tolerate agricultural biotechnology company Monsanto's Round-up herbicides, making it easy to eradicate weeds but tying farmers to using Round-up. Between 1997 and 2005, the total surface area of land cultivated with genetically modified crops rose from 17,000 km² to 900,000 km². Traditionally, farmers kept seeds from one crop to sow the next year. However, many GM crops do not produce viable seeds, closely linking farmers to suppliers.

Although the globalisation of farming has meant we have become capable of growing vast quantities of food, two downsides have been the continued narrowing of plant foods available and loss of biodiversity from land clearance and other human-induced pressures. Today, in the West, we now rely primarily on 12 plant species for 80 per cent of our calorie intake. More than half of the world's food energy comes from a limited number of varieties of three mega-crops: rice, wheat and maize. Creating plants with traits useful to farming on a large scale have sometimes resulted in the removal of useful characteristics, such as those that provide resistance to pests or tolerance to different climatic regimes. We are relying on just a handful of plant species for our survival and with climate change beginning to affect the world, the varieties we use may not be resilient enough to tolerate forecast rises in temperature and sea level and the accompanying shifts in pests and diseases.

Biodiversity is important for our survival as ecosystems that are intact provide us with natural services such as cleaning air and water, and regulating floods. Today, nations are recognising that farming using today's globalised methods cannot continue without Earth's ecosystems breaking down. With the world's population forecast to rise from 6.8 billion to 9.2 billion by 2050, we need to discover food production methods that help maintain or restore ecosystems, use water efficiently and do not cause huge emissions of greenhouse gases. A switch to better adapted varieties, agriculture that mimics and supports natural systems, and growing food at smaller scales for local communities could contribute to achieving this. As well as being good for the planet, it may well be good for us. With only 200 generations having elapsed since the beginning of farming in Europe, some experts believe humans remain better adapted to a diverse wild plant-based diet than the more limited agrarian one nearly all of us consume today.

ABOVE: An oil palm plantation with undergrowth of ferns in Cameroon. Palm oil is produced on a vast scale to fulfil our need for cheap oil in food and detergents. Rich rainforest biodiversity has been destroyed to make way for the plantations.

TSODILO DAISY

BOTSWANA: The north-west corner of the country, about 53 km from Shakawe. Three hills rise abruptly from the dry, bush-covered landscape. These are the Tsodilo Hills, sacred to the San Bushmen, who have, over thousands of years, painted images on the rocks, creating one of the world's most important collections of rock art. But the team from the Millennium Seed Bank Project in Botswana is searching for a very different kind of treasure: an extremely rare plant.

They are looking for the Tsodilo daisy, *Erlangea remifolia*. It has been seen in this vicinity before, but there are only about 50 individual plants left in the wild and the team has no idea whether they will find the plant, and if they do whether there will be seeds that they can collect to help improve its chances of survival.

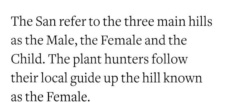

The San refer to the three main hills as the Male, the Female and the Child. The plant hunters follow their local guide up the hill known as the Female.

Plant hunting and seed collection in Botswana is a difficult and risky business with threats from wild animals, including lions, buffalo, scorpions and snakes to contend with. And hunting for rare plants in this landscape is like looking for the proverbial needle.

ABOVE Various views of the Tsodilo Hills.

While searching the hill, 300 metres high, they encounter a small puff adder and a large black mamba, one of Africa's most venomous snakes. But they go on criss-crossing the rocks carefully in the hope of finding their goal: a plant that grows only in Botswana and is therefore of national importance.

They search for hours and are beginning to give up hope when one of the team looks up and spots something purple on a cliff-face ledge, high above.

THIS PAGE: The hunt for the Tsodilo daisy in action.

It takes a bit of mountaineering, but the seed collectors eventually reach the ledge and are delighted to discover the Tsodilo daisy in purple flower. Best of all, there are seeds of this endangered plant. Fortunately, the ledge on which the plant grows is out of sight of tourists, who come to view the rock art.

The team collect some seed and head back to their camp, exhilarated that the expedition is such a success. Finding a Red Data Listed species in seed is like winning the lottery.

Seed is now safely banked in Botswana and in the UK.

Plant profile

COMMON NAME: Tsodilo daisy

LATIN NAME: Erlangea remifolia

FAMILY: Asteraceae

STATUS: Red Data Listed

POPULATION: Fewer than 50 or so individual plants

SIZE: 20-100 cm tall

DESCRIPTION: Long-lived annual or perennial herbaceous subshrub with purple flowers

THREATS: Tourism and development

Banking wild seeds gives an insurance policy for plants

The world's wild flora is in a sorry state. Of the 400,000 plant species that scientists estimate exist on Earth, as many as a quarter are threatened with extinction. The conversion of land for agriculture and urban development, plus unsustainable exploitation of wild plant stocks, are taking their toll. Every year, an area the size of England is cleared of primary vegetation. A constraint to conserving intact native vegetation is that scientists don't know enough about many species' status to gain protection for them through legislation.

Botanists are still unearthing new species at a rate of 2,000 a year and many plants that are known to exist grow in remote or inaccessible places that makes studying them in detail very difficult. What we do know is that wild plants are vital to humans as food, fuel, clothing, building materials and medicines. Many poorly studied plants could yield new nutritious foods, valuable biofuels or cures for diseases. Conserving existing plants and seeking out new ones is therefore of the utmost priority.

'Many wild species are already useful and many are potentially useful,' explains Paul Smith. 'For example, there are 30,000 species of edible plants that are not domesticated in the true sense but which give us the ability to innovate in the future. That will be essential if we are going to adapt to climate change and overcome food security issues caused by overpopulation. It doesn't just apply to food but to all utilitarian species, including those that provide vital ecosystem services. The best way to preserve biodiversity is to conserve, protect and manage plants *in situ*, in their own habitats where they interact with a multitude of other organisms. However, *ex situ* conservation away from their natural habitats gives a range of other options,

of which seed banking is one. It is the most efficient method for storing plant diversity, but at some point you need to grow the plants so understanding germination is key. Here at the Millennium Seed Bank Partnership we believe there is no technological reason why any plant species should become extinct.'

When scientists began realising that we were destroying Earth's resources, they convened the first international Earth Summit in Rio de Janeiro, Brazil, in 1992. The heads of state in attendance signed documents outlining policies for achieving sustainable development to meet the needs of the poor and recognise the limits of development to meet global needs. Among the treaties was the Convention on Biological Diversity (CBD), which aims to conserve biodiversity, promote the sustainable use of plants and animals, and ensure that the benefits arising from genetic resources are shared out equally. The Global Strategy for Plant Conservation (GSPC), an action plan for plants, grew from the CBD. This lays down 16 targets that include: compiling a list of known plant species as a step towards creating a world flora; conserving at least ten per cent of the world's ecological regions; and ensuring that 30 per cent of plant-based products come from sustainable sources. As well as helping develop the GSPC in 2002, the Royal Botanic Gardens, Kew, has placed the GSPC targets at the heart of its plans for the future. The MSBP will play a central role in helping it meet those goals.

There are plenty of good reasons why we need to conserve wild plants. For a start, many people around the world still get their primary nutrition by harvesting plants directly from the wild. Studies conducted in countries of sub-Saharan Africa indicate

that between 25 and 40 per cent of people get their food directly from nature. In developing countries, livestock also generally eat wild species. Enabling farmers to cultivate important food plants for their subsistence and to sell in local markets can help relieve pressures on wild plant stocks, as well as reducing poverty within rural communities. It is often lack of information and access to seeds that prevent farmers from cultivating plants rather than harvesting them from the wild. Before the MSBP places any new species into storage, its experts work out how to germinate the seeds. Passing this information on to people who can use it to help conserve wild plants is an important part of the MSBP's work. 'By simply delivering germination protocols to people in sub-Saharan African we have been able to help farmers cultivate fruit trees that they were previously only able to harvest from the wild,' says Paul.

In the West, we have considerably narrowed down the range of farmed plants on which we rely. Centuries of cultivation mean that our staple food crops have become genetically much less diverse than their wild ancestors. There are numerous cases in history where genetically limited crops have succumbed to stresses from pests or climatic changes, with disastrous results. The potato famine of the 1840s was caused by large-scale crop failures as a result of genetic vulnerability to the late potato blight epidemic. A large proportion of susceptible potato varieties grown at that time were wiped out as the blight spread across Ireland, continental Europe and North America. One million people died and a further million were forced to emigrate as a result. The southern corn blight outbreak in the US maize crop in the 1970s, together with large-scale rice losses in the

Philippines and Indonesia, have further highlighted the risks of relying on a few high-yielding varieties.

The wild relatives of modern-day crops offer a means to boost the genetic diversity within cultivated plants. In recent years, crop-wild relatives have been used to improve the resistance of crops to pests and diseases including late blight in potato, wheat curl mite and grassy stunt in rice. They can also improve tolerance to stressful conditions such as drought and salinity. The MSBP has so far sent 154 collections of salt-tolerant pasture species for use in trials in Australia, where over 5.7 million hectares of agricultural land are affected by salination. With climate-induced sea level rises forecast to inundate vast low-lying areas currently used to produce crops, the race is on to find more useful plants that can tolerate salt water. 'The problem is that when we breed plants for efficiency we sometimes breed out useful traits such as disease resistance, drought tolerance and so on,' says Paul. 'If we want to cultivate plants on marginal lands we need those genes to breed them into crop varieties for the future. We're working on a pilot project with the Global Crop Diversity Trust. We will be collecting crop wild relatives of 23 important crop species, which have specific traits such as drought tolerance and disease-resistance, so they can be used in breeding programmes.'

Aside from food, Chinese and Ayurvedic systems of medicine are relied on by millions of people, not just in Asia but increasingly around the globe. Practitioners historically relied on wild supplies but over-exploitation now means many of these useful plants are becoming rare. In China, the third most biodiverse country in the world, as many as 50,000 plant species are collected for

traditional medicine. Pressures from this industry, along with intensification of agriculture, large land developments and overexploitation of timber are putting China's biodiversity at risk. Scientists from the MSBP have partnered with the Chinese Academy of Sciences and helped establish a new seed bank, the Germplasm Bank of Wild Species at the Kunming Institute of Botany, to safeguard China's flora. 'The majority of our medicines originated from plants,' says Paul. 'The seeds we store could provide a cure for cancer or a new source of Vitamin C for African children. It's not an exaggeration to say they could represent the difference between life and death for millions of people.'

A new use for stored seeds is emerging as nations realise the present and future benefits of having biodiverse ecosystems and intact habitats. Called restoration ecology, it involves attempting to return damaged habitats to their natural states. Healthy ecosystems provide valuable services such as cleaning air and water, as well as protecting cities from floods. Without them, we lessen our chances of reducing poverty, hunger and disease. The MSBP and its partners are presently involved in habitat rehabilitation projects to restore forests in Kenya, pastures in Burkina Faso, and the natural habitat and favoured food plants of the endangered Regent honeyeater (*Xanthomyza phrygia*) in Australia. As climate change places new pressures on the globe's beleaguered flora, restoring wild vegetation has the additional benefit of increasing carbon uptake by plants. Seeds banked by the MSBP offer not only the means to safeguard rare, useful and threatened species but to use those resources to nurture future populations and reverse the environmental degradation that has left planet Earth in such a woeful state.

SEED BANK FACT
Paying peanuts saves species

It costs £2,000 to secure the safe storage of a single plant species in the MSB vault. This is a small price to pay to safeguard something whose future value to humankind could be priceless. One third of the seed collections presently held in the seed bank are of economic value to people in their native country. The MSB has sufficient space to store more than half of the world's seed-bearing species.

ABOVE LEFT: **Urban development in Madagascar's capital city, Antananarivo.**
ABOVE: **Planting gold: A Brazilian subsistence farmer and his daughter work together sowing crops in the rocky soil.**

Wolfgang Stuppy

Seed Morphologist

Wolfgang Stuppy's job is unique. As Seed Morphologist for the MSBP, he selects, describes and photographs seeds that come from all over the world for safe storage in the seed bank at Wakehurst Place. He got the job in 2002 after gaining a PhD in Systematic Seed Anatomy and Morphology in Germany and working as a Threatened Plants Officer with the living collections at Kew. 'I don't know any other botanists whose work is one hundred per cent dedicated to studying the morphology of seeds,' says Wolfgang. 'Most people research the molecular make-up of plants these days; morphology is just not *en vogue* any more. But in the context of the MSBP, it makes a lot of sense to examine and document the external and internal structure of seeds. This kind of information is not only useful for identifying seeds but also allows us to predict how seeds might germinate and respond to being stored in the seed bank.'

On a typical day at the MSBP, Wolfgang or his assistant profile five different seeds. They first assess the external characteristics, such as size, shape, colour and any appendages, then cut the seeds open and look at their innards. After describing them, they photograph them, through a stereo- or scanning-electron microscope if the seeds are miniscule. With new seed collections arriving almost every week, Wolfgang has to be very selective about which seeds are described. He chooses those that broadly represent the plants in the MSBP, as well as anything particularly interesting or rare. The information produced is filed under the seed morphology category of the MSBP's Seed Information Database, which is accessible to researchers around the world via the internet. 'People working within seed banks often have to scarify seeds before germinating them,' explains Wolfgang. 'By using our images to study the internal structure of the seeds, they can avoid harming the embryo when they make cuts.'

Wolfgang also frequently fields queries from outsiders wanting seeds identified. On one occasion, a defending QC wanted to know whether it would be easy to identify cannabis seeds or whether it would require a specialist. Wolfgang was able to tell her that as cannabis is 'monotypic', having only one species in the genus, its seeds are unique. Members of the public also often turn to the MSBP to identify unusual finds from the beaches of Devon and Cornwall. These are often drift seeds, or 'sea beans', as collectors call them, of tropical plants from Central America that have travelled north on the Gulf Stream. Although profiling and identifying seeds remains an important part of Wolfgang's work, he increasingly works as a spokesperson for the MSBP. For example, when architect Thomas Heatherwick chose to highlight biodiversity by constructing the British Pavilion at the 2010 World Expo in Shanghai from 60,000 acrylic seed-bearing rods, Wolfgang flew to China to highlight the MSBP's role in sourcing those seeds.

Other promotional work has involved Wolfgang appearing on Radio 4's Midweek programme, giving talks as part of the Royal Institution Christmas Lectures and writing books. Along with Rob Kesseler, Wolfgang has written three books with the aim of sharing his enthusiasm for seeds with others. *Seeds: time capsules of life* and *Fruit: edible, inedible, incredible* contain detailed scientific information along with close-up images showing the array of colours, shapes and sizes that seeds come in. The third book, *The bizarre and incredible world of plants*, co-authored with Rob Kesseler and Kew palynologist Madeline Harley, is more accessible to lay audiences. 'Seeds are far from boring,'

AMAZING SEED FACT
All at sea about seeds

Historians studying the ship the *Edwin Fox*, the ninth oldest ship in the world, encountered shells of unknown seeds. Built in 1853 in India, the ship sailed the high seas for 20 years carrying sugarcane slaves to Cuba, cargoes of pale ale, and Crimean troops reputedly accompanied by Florence Nightingale. After the ship was towed to Picton, New Zealand, in 1897, moves began to turn it into a tourist attraction. The Edwin Fox Society approached Wolfgang to see if he could shed any light on what plants might have yielded the seeds and why they might have been on board. He was able to tell them that they came from the *Canarium* genus of subtropical and tropical trees. The fruits contain protein, fat and carbohydrate, making it an ideal food for sailors, while the seeds shells would have served a useful purpose as ballast.

explains Wolfgang. 'Hundreds of millions of years of evolution have gone into making them perfect. We want to show people how beautiful seeds are, so that they want to protect them. The books are a bit like Trojan horses but in good ways; people buy them for the beautiful images and because they want to know about seeds or fruits but in the end they are also taught about the conservation work of the MSBP.'

The sheer diversity of seeds inspires Wolfgang in his work. Although he specialises in seeds, he has to have a very broad knowledge of plants because he never knows what he will encounter next. 'I look at seeds from angiosperms and gymnosperms from all over the world,' he says. 'I never stop learning because I see new things every day. Plants are truly amazing. Unlike animals, they have the remarkable ability to use sunlight to make sugar from just water and carbon dioxide. In doing so, they not only produce their own food but also feed – either directly or indirectly – all animals and people on Earth. Plants play an important role in our lives in so many different ways, but because they seem static and make no sound, most of us do not think of them as true living beings. However, I feel they are fellow living beings with whom we share this amazing planet and, as such, they deserve more respect. After all, the continued existence of humans on this planet depends entirely on plants.'

OPPOSITE: **Wolfgang Stuppy, the MSBP's Seed Morphologist.**
ABOVE: **Empty shells of *Canarium* sp.**

Chapter 2

Conserving wild plants on a global scale

An expedition heads into the Aberdane Mountains, Kenya, to seek and bring back seeds.

How Kew Gardens became a powerhouse of plant science

The Royal Botanic Gardens, Kew, came into being in 1759, when Princess Augusta created a garden to contain 'all the plants known on Earth' on her Thames-side estate. Today, locals and tourists come each year in their thousands to admire the largest living collection of plants anywhere. Numbering some 30,000, these include: a Chilean wine palm (*Jubaea chilensis*) believed to be the world's largest glasshouse plant; an old cycad (*Encephalartos altensteinii*) brought back from South Africa by Kew collector Francis Masson in 1775; one of the first Wollemi pines (*Wollemia nobilis*) to be planted following the discovery of this ancient tree in Australia in 1994; plus a maidenhair tree (*Ginkgo biloba*) planted while the Princess was still wandering the Gardens in 1762.

But Kew's plants provide much more than a changing seasonal spectacle for garden enthusiasts. For 250 years they have underpinned the vital botanical, scientific and environmental research that has made Kew a powerhouse of plant-based knowledge and expertise. Today, 700 staff work at Kew in roles that span: naming and classifying plants; curating botanical books, papers and illustrations; investigating chemical properties of plants that might yield valuable medicines or cosmetics; looking after Kew's heritage buildings such as the iconic glass Palm House and the Chinese-style Pagoda; learning how to germinate and propagate seeds so that plants can be saved from extinction; and educating visitors and scientists, young and old, as to why plants are vital to our lives.

In Kew's early days, when large areas of the world were still relatively unknown, its aim was simply to collect unknown plants. For example, Richard Spruce contributed plants from South America, Joseph Hooker from the Himalayas and Ernest Wilson from China in the 19th and early 20th centuries. During colonial times, Kew's botanists became adept at identifying species that might be of economic value, acquiring seeds and setting up plantations in British colonies with suitable climates. Kew was instrumental in establishing the global rubber industry, for example, in South-East Asia. It set up off-shoot gardens around the world to help manage the movement of plants between different countries. In the early 20th century, some 160 Kew-trained scientists were working in botanic gardens in Asia, Africa, Australia and the USA.

At that time, Europe's colonial powers were all involved in transferring plants around the world. Spurred on by the desire for economic dominance, they gave little thought to what detrimental effects their actions might have on natural biodiversity, nor why it might be important. However, as far back as 1799, Baron Alexander von Humboldt expressed concern about the rate of felling of cinchona trees in South America and Richard Spruce concluded, 50 years later, that supplies of wild plants used as commodities would run out unless they were cultivated. After the end of the Second World War, concern grew about the impact humans were having on the environment. Following the launch of the International Union for the Conservation of Nature (IUCN), the first ever Red List of threatened plants was drawn up in 1970. It concluded that 20,000 plant species needed protection to ensure their survival.

From the 1970s, scientists began realising that saving individual species was not enough. Botanist and ecologist Ghillean Prance conducted pioneering research into the inter-dependencies between plants.

He demonstrated, after frequent expeditions to the Amazon, that harvests of the valuable wild Brazil nut could only be successful if the surrounding rainforest was healthy. This is because the tree requires female Euglossine bees to pollinate it, and they will only mate with males who successfully gather a cocktail of scents from several orchid species, all of which grow only in undisturbed forest. Ghillean also recognised the damaging impact the 5,500 km trans-Amazonian Highway would have on the forest. After he became Kew's Director in 1988, he inspired the organisation to focus on conservation.

Today, Kew's mission is to 'inspire and deliver science-based plant conservation worldwide, enhancing the quality of life'. It is directed by the Global Strategy for Plant Conservation (GSPC), the overall aim of which is to halt the current and continuing loss of plant diversity. A 2007 review of the GSPC noted that it had been successful in allowing botanical gardens to engage in the

work of the Convention of Biological Diversity. Kew's vast living and preserved plant collections, extensive library resources and botanical artefacts, long heritage of studying plants from around the world and well-established relationships with foreign scientific institutions put it in a unique position to help deliver the GSPC's targets. As concern grows about the impacts of habitat loss, decreasing plant diversity and climate change, Kew is using its resources and expertise to identify species and ascertain how threatened they are, monitor ecosystems, influence legislation on biodiversity and restore damaged habitats.

Kew's resources underpin the MSBP's work conserving the world's flora. With its seven million specimens and numerous skilled taxonomists, the Herbarium is invaluable in helping MSBP collectors target plants to gather seeds from, classify new species found and ensure collections are correctly named. Data from Kew's Herbarium and that of

ABOVE: Kew's Palm House, an icon of this centre for botanical expertise.

France's Muséum national d'Histoire naturelle has been used to create collection guides for MSBP partner organisations. These provide information on what target plants look like, where they are likely to be found, the time of year and conditions under which they produce flowers and seeds, plus their conservation status. In the MSBP's first ten years, its collectors discovered a number of species that were new to science. Among the finds from Madagascar was an entirely new genus of Rubiaceae, the family to which coffee, gardenia and cinchona belong.

Kew's herbarium resources are used by the Geographic Information Systems Unit to help create digital maps showing exactly where target species might be found. The team overlay environmental data such as geology, climate and soils with known information about a given species' distribution, to highlight favourable locations. Such digital maps helped MSBP staff locate sought-after plants in Madagascar,

for example. Once at a collecting location, MSBP seed collectors record the names of plants they gather seeds from, take exact coordinates of the populations and note local environmental conditions. These data are later fed back into a geographical information system to help produce more up-to-date maps. Information from all MSBP seed collection sites are used to map global distributions of species and could be used in future to document the effects of climate change on these species.

Although gathering and storing seeds is fundamental to the MSBP's work, its underlying aim is to use those seeds for conservation and sustainability. This aim is being fulfilled, for example, through its work with Kew's UK Overseas Territories (UKOT) teams. The UKOTs comprise 16 former colonies that have elected to keep their direct British links and therefore form part of the nation state of the UK. Apart from Gibraltar and the British Antarctic Territory, they are all small islands.

Because of their isolation, islands contain particularly high numbers of unique species but many are at risk from over-exploitation, habitat loss and invasive species. The MSBP so far holds seeds from 270 UKOT species. It is working with its local partners to gather more and with colleagues at Kew to help restore damaged native habitats.

One flora that Kew and the MSBP are working to rehabilitate is that of the Caribbean island of Montserrat. Large tracts of the island's pristine cloud forests were destroyed following repeated eruptions by the Soufrière Hills volcano from 2005. The island has around 800 native species, including three that are unique to the island: *Xylosma serratum*, a member of the willow family (Salicaceae) that is now considered extinct; the orchid *Epidendrum montserratense* and a member of the coffee family (Rubiaceae) *Rondeletia buxifolia*. Until 2006, the latter species was only known from a book about Montserrat's flora. However, during fieldwork to assess vegetation in the Centre Hills region, Kew botanists rediscovered populations of the critically endangered species. MSBP staff subsequently conducted tests to work out the best conditions for germinating and growing the shrub. The Botanic Garden in Montserrat has now grown a demonstration hedge to assess its potential to replace introduced species currently used for hedging. Seeds of both *Rondeletia buxifolia* and *Epidendrum montserratense* are now in safe storage at the MSBP.

In future, Kew aims to expand its restoration ecology work in line with its Breathing Planet programme. A response to the 2006 Stern Report's conclusion that the cost of doing nothing about climate change is much greater than the one per cent of GDP needed for

effective action, the programme lays down seven ways in which Kew plans to use its unrivalled botanical resources and knowledge to ensure we use Earth's natural resources more sustainably in future. The goals include safeguarding 25 per cent of species by 2020 through the MSBP, building a global network to restore habitats, growing locally appropriate species for a changing world and assisting conservation efforts on the ground. Two and a half centuries on, the garden Princess Augusta created to contain 'all the plants known on Earth' is helping ensure the world's floral riches will not be consigned to contrived botanical collections but allowed to flourish in their natural habitats around the world.

Reaching out to save seeds and halt biodiversity loss

The Millennium Seed Bank Partnership is a response to the demise of our wild plant populations. The pressures of deforestation, urban and agricultural development and climate change are robbing the Earth of its remarkable biodiversity before scientists have had a chance to catalogue the estimated 400,000 species of plants that exist in its savannas, rainforests and deserts. This is bad news given that we rely on plants and biodiversity for food, building materials, clothing, medicines and for a range of cleansing environmental services. The chances are that there are plants going extinct out there that might have otherwise provided valuable new crops, remedies or fibres.

Scientists and researchers began realising all was not well with the world as far back as the late 19th century, when plant hunters noticed the destructive impact caused by overexploitation of wild plant commodities such as cinchona, a malaria treatment. Since then, our awareness of our planet's diminishing health has grown, thanks to the development of technologies such as remote sensing and Google Earth. These have improved researchers' abilities to map and monitor distributions of different plant species and revealed a global trend that some scientists are calling the world's sixth great extinction, on a scale comparable to the extinction of the dinosaurs 65 million years ago. Scientists estimate that between 150 and 200 species of life become extinct every 24 hours.

By storing seeds and learning how to germinate them, the MSBP aims to provide an insurance policy against imminent and future plant extinctions and to reverse the ongoing degradation of biodiversity by helping communities cultivate plants rather than exploiting wild stocks. Contained within the MSB's subterranean vault,

in the grounds of Wakehurst Place, the 400-year-old estate that Kew leases from the National Trust, the MSBP has so far banked seeds from more than 30,000 species. Among these are seeds from practically all of the UK's flora, and ten per cent of the world's flora. It now aims to safeguard 25 per cent of the world's plant species by 2020.

Seeds have an important place in the development of the human race. The transition our ancestors made from being hunter-gatherers to farmers involved them domesticating wild plants as crops. By favouring plants whose seeds germinated quickly, and which fertilised themselves, early farmers were able to grow reliable food crops with relative ease. While this enabled the global population to swell, it did so to the detriment of genetic variability that enables wild plants to adapt to changing environments. Concerns about the loss of locally adapted varieties in centres of high crop diversity prompted Soviet scientist Nikolai Vavilov to open the first seed bank in Leningrad in the 1930s.

When the United Nations convened the first Conference on the Human Environment in Sweden in the 1970s, the then Kew Director, John Heslop-Harrison, set about developing the Gardens' traditional seed-exchange programme, involving annual collections of seeds for exchange with other botanic gardens, into a network of wild seed banks. Some scientists were concerned that domestication of plants had changed the storage and germination behaviour of their seeds, so Kew wanted to conduct experiments to find out. This involved moving beyond simply collecting seeds from plants grown in the safe haven of a botanic garden to gathering and experimenting with those from plants growing in the wild. Collection trips began with a visit to the Mediterranean in 1974.

By the end of the 1980s, scientists in Kew's physiology department had demonstrated that seeds from wild plants could be stored in dry conditions at low temperatures for several years, in much the same way as crop seeds. Over a decade and a half, they had collected and conserved seeds from 4,500 species, in collaboration with organisations in the plants' host countries. By now, global concern for biodiversity loss was growing as natural history TV programmes such as those presented by Sir David Attenborough drew the public's attention to the issue. In 1992, at the first international Earth Summit, nations began signing up to the United Nations Convention on Biological Diversity. Setting out targets to halt and reverse biodiversity loss, it now has 193 Parties.

By the early 1990s, Kew was seeking once more to expand its seed-banking operations but needed external funding to do so. The team involved in seeking finance discovered that the National Lottery was to support five good causes, including a national celebration of the changing millennium in the year 2000. After a lengthy application process, the Millennium Commission awarded the MSBP a grant of £30million; corporate donations from Orange plc, the Wellcome Trust and others added to the pot. In 2000, the MSBP's futuristic building opened, with storage freezers capable of holding seeds from at least 50 per cent of the world's flora.

'That money allowed us to develop a key plan of action for studies to better understand seed diversity, how seeds survive dehydration, whether they survive freezing, what the optimum temperature is for sub-zero storage, why some seeds don't survive dehydration and what we might do about those,' explains Hugh Pritchard, Head of Research within Kew's Seed Conservation Department. 'We've also been able to carry out more fundamental studies in molecular biology, biochemistry and biophysics. For example, we are examining similarities between how seeds die and how human cells die.

OPPOSITE: The Montserrat orchid *Epidendrum montserratense* is restricted to a small area on the island of Montserrat. Much of its natural habitat has been destroyed or damaged, but two collections of seed are now safely stored in Kew's Millennium Seed Bank.
ABOVE: The MSB's Orange Room, with water plant display in the foreground.

This science will help us predict the potential storage behaviour of 300,000 flowering plant species.'

When the building opened, the team had already collected seeds from practically all the UK's wild plant species and, a decade later, as the first phase of the project drew to a close, it achieved the target of gathering seeds from ten per cent of the global flora. Central to achieving this target has been its policy of forming mutually beneficial partnerships in countries targeted for seed collection and operating its collecting programmes in a way that supports the letter and spirit of the Convention on Biological Diversity. Today, the project is the largest *ex-situ* plant conservation project in the world, working to conserve seeds and global biodiversity with partners in more than 50 countries.

The MSBP is now entering Phase II, a progression reflected in the name change from Project to Partnership. This phase runs from 2010 to 2020, during which time it aims to expand its collection to include seeds from 25 per cent of the world's flora. Initially viewed primarily as a 'Doomsday' insurance policy for plants, the bank is now making efforts to shift the emphasis from simply storing seeds for posterity towards putting its seeds and expertise to good use. As well as sending out seeds in response to requests from external users, it is considering using some of its stock to grow up nurseries of plants to produce more seed stocks that could be sold on. And, although the MSBP never uses seeds for commercial purposes, it is investigating ways that it can put its expertise and networks to good sustainable use.

A recent request to Kew for help with a project exemplifies the work it supports. Desterio Nyamonga from the Gene Bank of Kenya undertook his PhD at Wakehurst Place on a plant genus called *Vernonia*. The oil in seeds of *Vernonia* species is often used in the paint industry. 'He approached us to see if we might be able to help him establish a sustainable paint business

in Kenya, using the *Vernonia* plants he studied for his PhD,' explains Tim Pearce, International Coordinator. 'We at Kew may be able to him find a commercial partner in Kenya, or we may be able to assist with product development or market research. More and more in future we have to be well placed to help our partners realise value. One way to achieve sustainability in the use of a genetic resource is to commercialise it. But what we cannot do is to take a path that leads to the development of *Vernonia* plantations in place of wild grasslands.'

Kew's Innovation Unit has been set up to find ways in which it can lend its expertise and resources to sustainable commercial enterprises. Kew is renowned for its independent scientific research, has links with more than 800 organisations in 100 countries, and actively contributes to the Global Strategy for Plant Conservation. It aims to use these strengths to help companies make a difference in managing and conserving biodiversity at locations around the globe. The skills it can offer companies range from advising on sustainable trade of CITES-controlled plants (Convention on International Trade in Endangered Species) to conducting vegetation surveys and propagating critically endangered species.

'When you look at potential commercial projects, initially there will be technical impediments,' explains Tim. 'These can be overcome by science and technology, which are Kew's core activities. Then companies will need to know how to grow a particular plant, which we can also offer expertise in. And once they've grown it, they will need to know how to extract the useful oil or chemicals from it. There will also be the need for market development, and we may be able to play a role here as well. My belief is that because we've got good relationships with our partners we've got a great opportunity to bring financial value to their natural resources but in a sustainable way.'

A short history of banking seeds

Storing seeds for food and agriculture has a long history. Archaeologists have found caches of seeds at Early Neolithic sites, including Gilgal in the Jordan Valley, and Çatalhöyük East in modern-day Turkey. The latter contained both crop and wild seeds. Ancient Egyptian tomb paintings show grain storage, and archaeologists have found grain stores throughout Europe, including Skara Brae, a Neolithic village in the Orkneys. But storing seeds to conserve biodiversity is a recent idea and one in which the MSBP has played a major role.

ABOVE: Skara Brae, Orkney, Scotland.
RIGHT: Svalbard Global Seed Vault.
OPPOSITE: Seed germination tests using fire and heat, being run at Wakehurst Place.

Samples of particular populations of seeds are known as accessions. Between them, the world's seed banks store millions of accessions; some are cultivated, crop or agricultural seeds and others are wild plant seeds. But this is a tiny fraction of the world's biodiversity and plant species unknown to science are still being discovered. For example, in 2009, the year Kew celebrated its 250th anniversary, its botanists discovered over 250 new plant species including spectacular palms, minute fungi and giant rainforest trees. The new species came from a range of locations including Brazil, Cameroon, East Africa, Madagascar, Borneo and New Guinea. Experts believe that almost a third are endangered.

THE N. I. VAVILOV INSTITUTE OF PLANT INDUSTRY

Founded as the Bureau of Applied Botany in 1894, the N. I. Vavilov All-Russian Scientific Research Institute of Plant Industry, St. Petersburg, (formerly Leningrad), houses one of the world's first seed banks. It still holds one of the world's largest collections of plant genetic material; over 320,000 accessions from 2,532 species and 425 genera.

Its seed collections were largely built by Nikolai Vavilov, inspired director from 1921 to 1940. He organised expeditions worldwide, to gather wild and cultivated corn, potato tubers, grains, beans, fruits and vegetable seeds, while developing his theory on the centres of origin of cultivated plants. Vavilov fell from favour and died in a Soviet prison in 1943. The seed bank was preserved, even through the Siege of Leningrad despite desperate food shortages; one scientific assistant reputedly starved to death surrounded by edible seeds.

THE SVALBARD GLOBAL SEED VAULT

Opened in February 2008, The Svalbard Global Seed Vault is operated by the Global Crop Diversity Trust and is a global effort to safeguard the biodiversity of crops important for food or agriculture.

Deep in the permafrost of a mountainside on the Norwegian island of Spitsbergen, three concrete chambers have the capacity to hold 3.5 million seed samples. Other major seed banks, such as the National Center for Genetic Resources Preservation in the US and the Vavilov Institute of Plant Industry in Russia are also devoted to banking seed of, and research into, agricultural species.

The history of seed banking at Kew

1 The first recorded work on scientific seed storage at Kew was in 1898, when H. T. Brown and F. Escombe published 'A note on the influence of very low temperatures on the germinative power of seeds'.

2 Even then, back in the 1890s, seed storage was not new to Kew; the horticultural staff had collected and stored seeds from the Gardens to exchange with botanical gardens around the world for many years. But the stored seeds were not always viable.

3 Great improvements in seed storage came about after Peter Thompson was appointed physiologist at Kew's Jodrell Laboratory in 1964, and worked with the horticultural staff. The collaboration marked the birth of an idea that grew to become the Millennium Seed Bank Partnership.

4 The next few years clarified ideas about the management and process of seed storage, and reliable identification and recording methods. Germination tests for viability led to a full time seed-handling and testing laboratory attached to a seed store.

5 By 1968, the building provided seed reception, handling, sorting and testing facilities for seed from the Gardens. Its potential for conserving seeds of endangered plants was first recognised when Kew agreed to store seed from plants threatened by a new reservoir in Upper Teesdale.

6 In 1973 the Physiology Section moved to Wakehurst Place and was joined a year later by the seed collection, which moved into two temporary cold stores in the chapel. Peter Thompson headed the new section and, during the early 1970s, the focus shifted to vegetation under threat.

7 Kew's first major seed collecting expedition went to the Mediterranean in 1974.

8 During the late 1970s and early 1980s development of the seed bank and the seed research programme continued and included a modest purpose-built seed bank and a computer for data handling.

9

In 1980, Roger Smith became head of the seed research and banking operation at Wakehurst Place. The enterprise was in the Mansion, the former home of Sir Henry Price and his family. Janet Terry, who arrived to work in seed banking in 1984, recalls, 'our main lab was in Lady Price's bedroom, which still contained a fur safe in the back of a cupboard. Seed cleaning was done in Lady Price's daughter's bedroom, and the X-ray machine was in Sir Henry's bathroom.'

10

The first full-time seed collector was appointed in 1989 followed by another in 1990 (supported by Marks & Spencer).

11

The idea of a global seed-banking project supported by a purpose built seed bank and research institution had been bubbling for years. It was conceived after the Rio Earth Summit in 1992 to try to ensure the survival of the estimated 60,000–100,000 seed-bearing plant species threatened by human development and climate change. In 1995, a proposal for the Millennium Seed Bank Project was submitted to the Millennium Commission. It was successful and, in 1996, HRH The Prince of Wales, with Sir David Attenborough as patron, launched the Millennium Seed Bank Appeal. Orange plc became its premier sponsor and in 1997 the Wellcome Trust announced a substantial donation to the MSBP. The construction of the Wellcome Trust Millennium Building, to house the project, began in 1998.

12

Conservation work continued alongside these major developments and 1997 saw the launch of the UK Flora Programme to collect the seeds of 1,400 native UK plants. It was substantially completed in three years, making the UK the first country in the world to have such a complete collection of its native flora.

13

On 20 November 2000, HRH the Prince of Wales opened the newly completed Wellcome Trust Millennium Building. The seed conservation project moved into a new and important phase involving partner organisations around the globe. Seven countries signed collaboration agreements in 2000, rising to 16 by 2003.

14

In 2004, the EU funded the European Seed Conservation Network (ENSCONET) for five years. The network involved 19 institutes from 12 countries including the Royal Botanic Gardens, Kew, which acted as coordinator.

15

Roger Smith OBE retired in 2005 and was succeeded by Paul Smith.

16

The project collected its billionth seed in 2007 and presented it to Gordon Brown (then Chancellor of the Exchequer).

17

By 2009, the MSBP, working with a network of partners across 50 countries, had collected 10 per cent of the world's threatened plant species. It was a major achievement: the seeds of 24,200 species collected 14 months ahead of schedule and under budget.

18

The next phase of the Millennium Seed Bank Project, known as the Millennium Seed Bank Partnership, is an even more ambitious target of collecting a quarter – about 75,000 – of the world's plant species threatened by human development and climate change, by 2020. The aim is also to increase the use of banked collections by organisations researching and delivering the sustainable use of plants and the restoration of damaged habitats.

Two types of seeds pose a quandary for conservation

BELOW RIGHT:

Swartzia seeds, an example
of recalcitrant seeds.

Botanists recognise two main types of seeds: orthodox and recalcitrant. Orthodox ones can tolerate desiccation, while recalcitrant ones cannot. As seeds must be frozen to undergo long-term storage, and have to be dried first to avoid ice destroying plant cells, only orthodox seeds can be easily banked. As a rule, orthodox seeds tend to come from dryer parts of the globe, where they have evolved over time 'to be dry but not die'. The MSBP and its partners therefore avoid collecting seeds from moist rainforests in favour of drylands or temperate ecosystems.

MSBP scientists have been trying to find unique characteristics that would indicate to collectors in the field whether seeds are orthodox or recalcitrant. Generally, smaller and flatter seeds are more tolerant of desiccation, while larger, more rounded ones are less tolerant. This is because small, flat seeds have higher surface area to volume ratio and therefore a greater chance of drying out.

'If you dry a tea towel by hanging it on the line in a single layer, then it will dry in two minutes flat,' explains Hugh Pritchard, Head of Research within Kew's Seed Conservation Department. 'But if you roll it up into a sausage shape, it's the same volume but takes longer to dry. By the same reasoning, if you are a seed and you may die from drying, it makes ecological sense to be big and not let your water go so quickly. Recalcitrant seeds also tend to have thinner seed coats. This is because they will germinate and reach any water in the soil more quickly with a seed coat that's easy to break through.'

Dried and frozen seeds of orthodox species can live for tens or even hundreds of years. In 2005, Israeli scientists successfully germinated a 2,000-year-old seed from a Judean date palm excavated from Herod the Great's palace in Masada, Israel. The oldest seeds that scientists in the MSBP have germinated were 200 years old (see pages 120–3).

No one knows how many of the world's flowering plants have orthodox seeds and how many have recalcitrant ones. The majority have orthodox seeds but, with the world's moist, tropical rainforests potentially containing 50 per cent of flowering species,

An inexact science

as many as a quarter of known plants could have recalcitrant seeds. The MSB is therefore also carrying out research to find ways to preserve recalcitrant seeds. For example, it is studying cryopreservation techniques, where tissues are cooled below the freezing point of water. One method involves using liquid nitrogen to preserve the embryonic axis part of seeds.

The MSBP is partnered with the Plant Germplasm Conservation Research Group (PGCR) at the University of Kwazulu Natal, pioneers in storage of recalcitrant seeds. Scientists from the two organisations joined forces to set up the Cryo-Conservation Centre of Excellence for sub-Saharan Africa, with the aim of conserving useful recalcitrant species. To date the project has screened the seeds of 69 species, identified 58 as recalcitrant and successfully cryopreserved more than half of them.

MSBP staff also worked on citrus cryopreservation with Griffith University, Australia, and are currently working on increasing understanding of Amazonian species with the Brazilian Institute on Amazon Research, Manaus. Many of the dominant trees in rainforests have recalcitrant seeds, so being able to preserve them has important implications for the timber industry as well as for global biodiversity.

Some seeds fall into a category somewhere between orthodox and recalcitrant. These are known as intermediate. It is also possible for seeds within the same species to show variation in their desiccation tolerance, linked to climate. MSBP scientists analysed seeds from the sycamore tree, *Acer pseudoplatanus*, in locations spanning 21 degrees of latitude across Europe. Trees growing within their native range, in France and Italy, received considerably more energy from the sun over the test period than did those in Scotland. Seeds from the former were more tolerant to drying and classified as 'not recalcitrant'; those from the latter were more desiccation-sensitive and classified as recalcitrant. 'Previously, seeds from this tree had been classified as recalcitrant,' says Hugh Pritchard, Head of Research within Kew's Seed Conservation Department. 'However, our work shows that there is a continuum in seed traits linked to the energy received during the growing season. The seeds harvested from northern latitudes are much more immature than those at southern latitudes. This finding has implications for modelling the impacts of climate change, as it shows what effect a temperature increase would have on seed development.'

The three 'E's provide a priority for saving

The MSBP's emphasis when collecting seeds is governed by the needs of partnering host nations. However, targeted species generally fall under the umbrella of the 'three 'e's': plants that are endangered, economically valuable or endemic (not found in the wild anywhere else). Between a third and a half of plants can be classified in this way. 'Our emphasis when collecting varies from country to country, depending on national priorities' says Paul Smith, Head of Seed Conservation at the Royal Botanic Gardens, Kew. 'In somewhere like Kenya or Burkina Faso we'd be collecting mostly useful species but in South Africa or Chile the focus would be on rare or threatened plants.'

The word endangered is used to classify species that are at a very high risk of extinction. The International Union for the Conservation of Nature (IUCN) assesses and classifies species as Least Concern, Near Threatened, Vulnerable, Endangered, Critically Endangered, Extinct in the Wild and Extinct. It publishes a Red List of Threatened Species, which is updated annually. In making its assessments, the IUCN draws on the knowledge of expert scientific institutions, including Kew. However, sometimes assessments are made using herbarium data on known distributions, often gathered in the long-distant past. As MSBP staff have found, these are not always reliable.

'We don't really have enough knowledge to tell just how many species are endangered,' says Paul. 'One of the great pluses of the MSBP is that we're out in the field gathering real data, not just extrapolating from herbarium distributions about the rarity of a species. In Botswana we started with a Red List of 13 species that were regarded as Vulnerable, Endangered or Critically Endangered. When we went out to look for those plants, only about three species were actually really threatened.

At the same time there are many species that are not on any Red List that are severely threatened. There's terrific uncertainty until you go out and look.'

In reaching its target of gathering seeds of ten per cent of the world's species, the MSBP has helped fill in many gaps in our knowledge of the conservation status of the world's flora. For each collection made, scientists record the locations of plant populations using GPS equipment and estimate population sizes; where possible this is used to update Red Lists within host nations. Many species that are definitely seriously threatened have been safe-guarded since the MSBP began in 2000. 'In South Africa *Erica verticilata* and *Erica margaritacea* were extinct in the wild but both have been reintroduced now through the work of the MSBP,' says Paul (see panel opposite).

The term endemic comes from the word *endemos*, meaning 'native' or 'in the people' in Greek. It is used in ecology to describe species that are exclusive to a particular geographical region. Because the distribution of endemic species is limited, they are

more vulnerable than other plants to habitat destruction. 'Some countries have very few endemics, for example neither the UK nor Botswana have very many. But somewhere like Madagascar is considered a hotspot for conservation because 80 to 90 per cent of the flora is endemic, which amounts to some 10,000 species.' On this large Indian Ocean island, the lack of protected areas and practice of 'slash-and-burn' agriculture are increasing the risk of extinctions.

Economic plants are those that can be used in some way by humans. Kew's Survey of Economic Plants for Arid and Semi-Arid Lands (SEPASAL) database lists more than 6,000 plants that people rely on for everything from stabilising land and growing hedges to contraceptives and dyes. 'These plants are not necessarily valuable on a global economics scale,' explains Paul. 'If a plant has a recorded use in the literature, it becomes a target for us. Also, we ask local communities which plants they use and what for and we target those species too. Around one third of the plants for which we hold seeds have a known use.'

The three 'E's exemplified

ENDANGERED

The MSBP helped bring back *Erica verticilata* from the brink of extinction. This attractive shrub, with whorls of tubular pink flowers, once grew on South Africa's Cape Peninsula. The last wild record of this erica was made in 1908, but sometime thereafter it became extinct in the wild.

In the 1980s, horticulturalist Deon Kotze of Kirstenbosch National Botanical Gardens began a concerted search for lost ericas. He was alerted to a plant growing at Protea Park in Pretoria, which turned out to be *E. verticillata*. Further specimens turned up over the years at various gardens, including Kew. Ted Oliver of the Compton Herbarium at Kirstenbosch confirmed that some of these plants were valid species.

Thanks to horticulturalists and MSBP partner staff at Kirstenbosch, this species is now growing once more at restoration sites near Cape Town, such as the Kenilworth Race Course Conservation Area, Rondevlei Nature Reserve and at Tokai Rehabilitation Project. In addition, seed has been collected and lodged in the Millennium Seed Bank vault.

ENDEMIC

Only 18 per cent of Madagascar's native vegetation remains intact and one third has disappeared since the 1970s. The MSBP's partnership with Madagascar is helping save large numbers of endemic species. So far, over five million seeds have been collected from more than 1,000 plant species.

ECONOMIC

Economic plants provide food, medicine, clothing, fuel and building materials for people all around the globe. One of the most useful of all is *Moringa oleifera*. Native to the sub-Himalayan tracts of India, Pakistan, Bangladesh and Afghanistan but widely cultivated across the tropics, the tree provides nutritious leaves and fruit, and is also a source of oil, cosmetics, lubrication, flocculant, fertiliser, dye and medicine. All 13 species of *Moringa* have outstanding economic potential. The MSBP team recently discovered a small wild population in Southwest Madagascar of *Moringa hildebrandtii*, which was thought to be extinct. The seeds are now safely stored and could be used to further research into this valuable genus.

STARFRUIT

BERKSHIRE, UK: It is June, a few hours after dawn and the air ruffles the surface of the shallow pond swaying the long stems and white flowers of the starfruit (*Damasonium alisma*), named after its star-shaped fruits. The flowers of this aquatic plant opened after dawn, but will last only for a day.

The MSBP holds seeds from four different UK sites. The original starfruit seed collections were small, but in a 2001 programme sponsored by Natural England, Kew horticulturalists based at Kew's Sussex site, Wakehurst Place, used seed to grow plants and harvested about 29,000 seeds.

The starfruit is so rare that coming across it is like finding treasure. It is listed as critically endangered and is at extremely high risk of extinction in the wild. The main causes of its decline are habitat loss and neglect: ponds in pastures where cattle once trampled the mud as they drank have largely disappeared and many other ponds are overgrown or neglected. But a remarkable collaboration to save it, between Plantlife, Natural England, West Berkshire (WBCC) and Buckinghamshire County Councils and the Royal Botanic Gardens, Kew, has been in progress since 2000.

THIS PAGE: Images of the endangered starfruit, *Damasonium alisma*.

The team get to work sowing starfruit seeds and seedlings.

On a November day, horticultural staff from Wakehurst and staff from the MSBP join representatives from Plantlife, Natural England and West Berkshire and Buckinghamshire County Councils at Greenham Common, Berkshire to try to secure the starfruit's future.

The team sow starfruit seeds and seedlings raised from MSB-stored seed into ponds on Greenham Common. Over 300 seeds and 60 seedlings are planted throughout the day. The hope is that starfruit will again thrive in wild ponds and go on to have an assured presence in the UK's flora.

THIS PAGE: The site of past protests is the perfect place to sow starfruit seeds.

Greenham Common was well known in the 1980s and 90s for protests against cruise missiles on the American Air Base, but following the departure of the Americans, missiles and all, a massive restoration programme is under way to help the site revert back to heathland.

It includes the introduction of grazing and, as part of the restoration, the creation of new habitats. Cattle regularly use the ponds on the Common and as they trample the margins they turn over the mud and bring starfruit seeds closer to the surface creating ideal conditions for them to germinate.

Plant profile

COMMON NAME: Starfruit

LATIN NAME: Damasonium alisma

FAMILY: Alismataceae

STATUS: Critically Endangered in UK; priority species UK Biodiversity Action Plan.

UK POPULATION: Very few sites in Buckinghamshire, Berkshire and Surrey

DESCRIPTION: Aquatic annual (occasionally perennial) with star-shaped fruits, small white flowers with yellow centres, blunt heart-shaped leaves that float when plant is submerged; flowers June to August

THREATS: Loss of habitat due to development and changing use of land; protected under Schedule 8 of the Wildlife Countryside Act 1981, which makes it an offence to intentionally pick, uproot or destroy any plants.

Paul Smith

Head of the Seed Conservation Unit

BELOW: **Dried and cleaned seeds of *Musa itinerans*.**
OPPOSITE: **Paul Smith showing the billionth seed (of the African bamboo *Oxytenanthera abyssinica*) safely banked in the MSB vaults.**

Paul Smith carries in his pocket a bottle of seeds from the mungongo tree. He uses it to demonstrate how the scientific work carried out in laboratories of the MSBP can translate to real benefits for people living in remote places such as Africa's Kalahari Desert. A member of the *Schinziophyton* genus that grows in Central Africa, the ten-metre-high mungongo bears fruits containing nuts rich in protein and vitamins. For centuries, African communities have harvested these seeds from the wild as a nutritional supplement. However, ongoing agricultural and urban developments have rendered the mungongo increasingly rare.

'There was no history of cultivating the species as locals didn't know how to germinate the seeds,' says Paul. 'Through the work of the Millennium Seed Bank Project we discovered that if you crack the nut and then treat it with smoke you get germination rates as high as 90 per cent. We passed that methodology on to tree seed centres in Africa, so this is now a species that people in Africa can cultivate. It's a wonderful example of how communities can now use something that had been inaccessible to them simply because of the missing knowledge on how to germinate it.'

Although now embroiled in the commercial and budgetary functions that heading Kew's Seed Conservation Department entails, Paul has ample knowledge of how plants are vital to rural communities. Having grown up in rural Africa, he later trained as a plant ecologist and spent three years undertaking a vegetation survey in Zambia. 'It was a fantastic opportunity,' he says. 'I was given a rifle, a Land Cruiser and an assistant and told to go and map 4,500 km² of national park in Zambia's Luangwa Valley. I learned the flora, which ultimately enabled me to get a job at Kew.'

After three years in the Herbarium transcribing the Zambian field notebooks of ecologist Colin Trapnell, and five years coordinating the southern Africa and Madagascar collections of the seed bank, Paul took up his current post in 2005. At the time, the MSBP was tracking somewhat below its collecting targets so he set about widening the seed-collecting programme. His efforts paid off when the MSBP met its target of collecting seeds from ten per cent of the world's flora by 2010 on time and under budget (see box opposite). Today, Paul's role is hugely varied. On a given day he

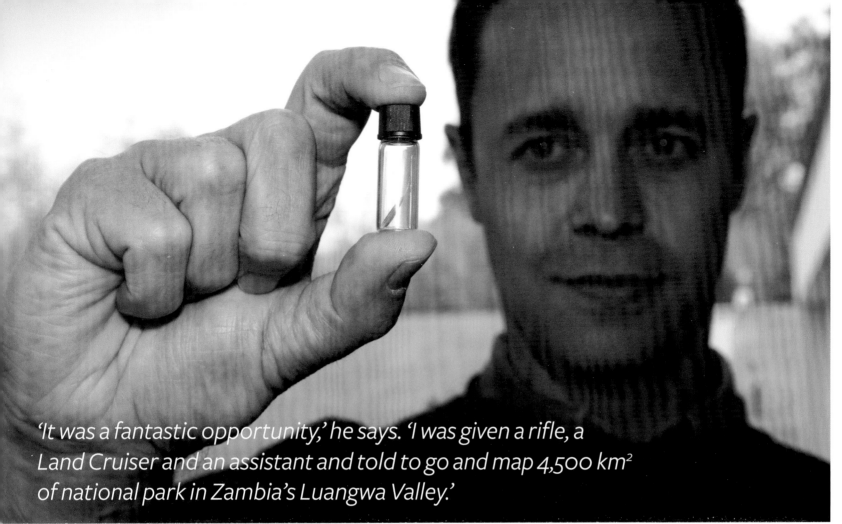

'It was a fantastic opportunity,' he says. 'I was given a rifle, a Land Cruiser and an assistant and told to go and map 4,500 km² of national park in Zambia's Luangwa Valley.'

might be working on plans for the post-2010 phase of the project (the Millennium Seed Bank Partnership), showing people around the seed bank, giving public lectures or visiting Kew Gardens in his capacity as Head of Department. He also finds time to visit the MSB's partners. 'Last year I'm afraid I won the prize for the largest carbon footprint in the department and was given a box of Odor Eater inner-soles as a booby prize. But the serious purpose was to meet with partners on every continent to celebrate past achievements and discuss what to work on in future,' he says.

Paul's vision for the future is for the MSBP to move away from the 'Doomsday vault' insurance model, oft-quoted by the media, and to look at how the MSBP's seeds can be used to ensure the long-term *in situ* conservation of species or enable people to use them. 'We're working on every single collection that comes in and finding novel ways to grow those plants,' he says. 'For me, our emphasis needs to be on adding to knowledge of the seeds we collect in a way that ensures their survival and enables our use of them. That research will also ultimately help secure our own survival.'

SEED BANK FACT

Seed saviours' top banana

The MSBP met its initial target of gathering seeds from one tenth of the world's species in 2010, when it received seeds from its 24,200th species. The plant achieving this accolade was the pink Yunnan banana (*Musa itinerans*) from China, a forest banana that is an important food for wild Asian elephants. A wild crop relative of commercially grown bananas, it is a valuable genetic resource that could potentially be used to breed new varieties with disease-resistance traits, thus ensuring the continued cultivation of bananas in the future. An official ten per cent banking ceremony, hosted by Kew's Director, Professor Stephen Hopper, and attended by the then Secretary of State for Agriculture Hilary Benn, took place at the MSBP on October 15th 2009. The ten per cent target was set in 2000 when the MSBP was formed. The MSBP now aims to collect and bank seeds from 25 per cent of the world's flora by 2020.

Chapter 3

In search of the world's seeds

Found at last: MSB International Coordinator Michiel van Slageren admires female cones of the ancient *Welwitschia mirabilis* in western Namibia.

Finding the best places to collect seeds

The MSBP collects seeds from all parts of the world, apart from tropical rainforests. The technique it employs to dry down and store seed at –20°C is applicable to orthodox seeds from anywhere, although research suggests that seeds from dryer places may remain viable for longer than those from wetter ones. 'We exclude tropical rainforests because expeditions to those locations require a good deal of research support, which is expensive, and there is more uncertainty about the number of seeds we could successfully store,' explains Michael Way, Head of the Collecting and Network Section. 'However, we can use the technique across most parts of the temperate world. The UK flora has been very successfully collected and stored, and the UK is by no means a tropical dryland. We're also using seed banking across the USA, southern Africa and Tasmania. We used to focus on drylands but collecting from places that are pretty cold and wet is also business as usual now.'

The range of terrains that MSBP seeds are gathered from is extensive, spanning the dense bush of south-west Kenya, fire-prone peaks and slopes in Western Australia, the rocky mountains of Lebanon, limestone karst topography of China's south-west province of Yunnan and the hummocky coastal dunes of Namibia. Physically accessing target sites can be difficult. Roads are often poorly maintained and prone to flooding or landslides; some locations are so remote the only means of access is by helicopter. Even if collectors actually manage to make it to a target site, weather, fire, animals and insects can all conspire to present additional challenges to collecting sought-after seeds. The rare plant *Acanthus syriacus*, which grows in the Middle East, is a case in point. It took the MSBP and its Lebanese partner organisation several attempts over four years to successfully gather seeds from the spiny Syrian bear's-breech (see pages 70–71).

Many of the MSBP's successes represent personal triumphs for their collectors, who frequently spend years tracking down lost, forgotten or favourite species. Tim Pearce has had a soft spot for large Afro-alpine plants since starting his career as a student in the late 1970s surveying the vegetation of the Aberdare Mountains in Kenya. Thirty-two years later, he and his Kenyan botanical partners went back to the location of his initial trip to collect seeds from a species of giant senecio, *Dendrosenecio battiscombei*. 'It's certainly been my best seed collecting experience to date,' Tim enthuses. 'When I first visited the site I was 18 and marvelled then at the sheer grandeur of the plant. It is a monster that looks a bit like a giant ragwort but with cabbage-like rosettes of leaves at the top of six foot, woody stems. These large Afro-alpine plants are real icons of botany. When we went back, there it was again in all its glory, a shower of bright yellow composite flowers. We got a fantastic seed collection. The site was just like it had been three decades before; I could remember exactly where I had camped and eaten my lunch.'

With intolerance to drying presenting the only technical restriction on what the MSBP can collect, the plants it safeguards come in all shapes and sizes. As of spring 2010, the banked species represented 340 of the 506 known plant families. They ranged from the UK's common daisy (*Bellis perennis*), which many would consider a useless weed, to the cream-flowered tulip *Tulipa bifloriformis*, a popular ornamental plant native to Central Asia, and the Madagascar periwinkle

ABOVE: **Abseiling in the Cederberg Mountains of the Westen Cape, South Africa, in search of the elusive *Protea cryophila*.**

OPPOSITE, LEFT: **The Madagascan *Catharanthus roseus*.**

OPPOSITE, RIGHT: **Collecting *Terminalia mollis* in Burkina Faso, West Africa, involves climbing and catching.**

Catharanthus roseus, compounds from which have been used to treat cancer. In all, the rows of glass jars in the MSB's 50 m² subterranean freezer room contain 1,654,753,608 seeds, representing about 30,000 saved species. Twelve of the species are now extinct in the wild. 'The chiller storing the seeds at the MSBP has the highest diversity of seed-bearing plants on the planet,' explains Simon Linington, Head of Management Support within Kew's Seed Conservation Department.

The fact that the MSBP now stores seeds from more than ten per cent of the world's flora is thanks primarily to the global network of official partnerships it has forged over the years. Just as the 16th-century botanist Carolus Clusius helped advance knowledge of tulips and other exotics unknown at that time by corresponding with some 300 collectors, patrons, botanists and apothecaries across Europe, so the MSBP has reached out and forged links with like-minded environmental institutes around the globe. As such, it is a leading example of how best to approach and help resolve global environmental problems. In the past, Kew accepted donations of seeds from researchers or botanists without necessarily knowing how they were collected or for what purpose. However, since the introduction of the Convention on Biological Diversity in 1992, Kew has only accepted seeds from officially recognised partners with whom it has signed agreements.

'We start out by finding institutes that have common objectives,' explains Michael Way. 'Then, in the early days of the partnership, we ensure there is very strong communication between us to make sure that expectations on both sides are properly brought in to formal legal agreements, work plans, budgets and

RIGHT: The eastern slope of Mount Lebanon above Ainata, with open annual vegetation of *Echinops* species (foreground left) and the endemic *Berberis libanotica* (right).

relationships that the partner has with local people, landowners, governments, permit authorities and so on. Communication and transparency are very important at the outset to make sure everyone knows what we're aiming to achieve. After that it's down to good project management. Successful partnerships derive from good objective setting and monitoring, plus active participation from Kew staff. We try to make sure someone attends a workshop, training event or expedition at least once a year. Then to maintain the partnerships, we put a lot of effort into producing reports and web pages to show what we've achieved.'

Michael cites the MSBP's partnership with the Autonomous National University of Mexico (UNAM) as an example of a good, mutually-beneficial, ongoing relationship. Kew's seed-collecting team had worked with the university on small seed-collecting and research projects for many years before the Millennium Seed Bank Project came into being in 2000. Then, in 2002, the MSBP signed an Access and Benefit-Sharing Agreement (ABSA) with the university at a high level, having checked in advance that the Wildlife Authority within the Environment Ministry was happy with the arrangement. 'We have spent the intervening years expanding seed collections through all the priority geographical regions in Mexico with a mix of contract and university labour supplied by UNAM and support from here,' explains Michael. 'We've also had an active research relationship with them. We've provided support and advice to their students doing seed studies, we had some of their staff come over to train with us and we've co-authored research papers. Recently, we've also started working

with some local Mexican communities to safeguard plants that are of use in their daily lives' (see pages 157).

Setting up and maintaining partnerships between the MSBP and far-flung institutes requires the odd skill set of being knowledgeable about plants, good at official diplomacy and capable of forging trusting personal relationships. International Coordinator Tim Pearce had already made lasting links with people at botanical organisations in Kenya, when the MSBP employed him to develop an official partnership with the country. These helped smooth the path when he was trying to negotiate an agreement that the Kenyan Government was happy with. Nonetheless the process took 18 months. Back in 2000, the Convention on Biological Diversity was relatively new and everyone

Useful but overused

The Kenyan native plant *Osyris lanceolata* has several uses. Essential oils are distilled to make perfumes; an extract from the bark is used to treat diarrhoea, chest problems and joint pains; and some communities use powdered bark from the plant as a substitute for tea. Unfortunately, commercial demand for products from the plant is now seriously threatening its existence within its natural habitat. Several conservation measures are now being put in place to safeguard this valuable plant, including banking its seeds for potential use in future ecosystem restoration programmes.

was beginning to feel their way round issues of access to genetic resources and sharing of benefits. Suspicions that the MSBP staff were nothing more than 'biopirates' were rife. 'One evening I met the Minster of the Environment and he turned to me and said, 'So you're the person who wants to steal our genetic resources',' recalls Tim. 'It was tongue-in-cheek but I'm glad those days are over.'

With the agreement for Kenya finally in place, the MSBP demonstrated its potential by uncovering several new species; promoting collaboration between formerly discrete foresters, agriculturalists and botanists; and using plants cultivated from gathered seeds to rehabilitate degraded sites. As the Millennium Seed Bank Project moves into its second decade,

becoming the Millennium Seed Bank Partnership, it aims to maintain successful partnerships in existing countries, as well as using its wealth of experience in negotiating mutually beneficial agreements to establish new fruitful links. 'We aim to prolong partnerships in very biodiverse places such as Mexico, where we have really only begun scratching the surface when it comes to exploring the flora, and to forge new relationships with countries such as Brazil and India, where we could also potentially have a lot of impact,' says Michael. 'Just this week we were contacted by a university in Mongolia. We are constantly evaluating approaches from potential partners and trying to work out where the best places are for us to deploy our skills.'

International partnerships

Malawi

Although nine per cent of Malawi's land is protected, deforestation and erosion are a menace outside these areas. The country has 5,500 species, 115 of which are endemic. The MSBP has been working with partners in Malawi since 2002. As well as helping to improve the country's seed-banking facilities, the MSBP has assisted with developing germination protocols and creating *ex situ* living collections of rare and threatened plants at the National Botanical Gardens.

SAFEGUARDED SPECIES: *Oldfieldia dactylophylla* is a rare fruit tree used by locals to cure skin diseases in cattle, as a treatment for diarrhoea in humans and to repel wild animals. Its native name *Nawonga* means 'thankful'.

South Africa

The Republic of South Africa's plains, hills, mountains and escarpments support forests, savannas, grasslands and the unique 'karoo' and 'fynbos' habitats. The country has 20,000 native species, thousands of which don't grow anywhere else and many of which are threatened by agriculture, urbanisation, drought and climate change. The MSBP provides services to a network of South African *in situ* and *ex situ* conservation initiatives, such as practical help with seed collecting, processing and germination, plus training.

SAFEGUARDED SPECIES: The Critically Endangered snow protea (*Protea cryophila*) only lives on rocky ledges on two of the highest peaks in the Cederberg wilderness area. The plants are usually covered by snow for months in winter. A collecting expedition in 2005 managed to obtain seeds, so these are now banked at the South African National Plant Genetic Resources Centre in Roodeplaat and at the MSBP.

Chile

The MSBP aims to conserve seeds from 20 per cent of endemic and 60 per cent of threatened species from the desert and Mediterranean ecosystems of Chile. Human developments and activities are increasingly destroying habitats in Chile, and existing *in situ* protection measures have had a limited impact. Seeds from over 700 native species have been conserved at the MSBP so far.

SAFEGUARDED SPECIES: *Dalea azurea* grows wild in a single valley in the Atacama Region in Chile. Seeds from this critically endangered plant have been banked in Chile and at the MSBP. These have been successfully propagated by seed and vegetatively, ensuring plants will be available in the future.

Burkina Faso

The MSBP's work in Burkina Faso is contributing to its goals of helping African communities secure sustainable plant resources and combat desertification. The MSBP supports Burkina Faso's Centre National de Semences Forestières by training staff, providing seed storage equipment, advising on seed collecting and helping restore degraded habitats within the semi-arid Sahel.

SAFEGUARDED SPECIES: the African mahogany (*Afzelia africana*) has been overexploited for timber and is now listed as Vulnerable by the IUCN. It is used in West Africa to make African drums ('djembe'), which are exported in large numbers. The wood is also used for construction, furniture, cooking utensils, canoes and charcoal. Its seeds and those of more than 1,000 other important wild plant species have so far been stored at the MSBP.

UK

The UK flora comprises around 1,400 species, of which more than 300 are threatened with national extinction. Agriculture, road-building, urbanisation, pollution and climate change are all putting pressure on wild plants. The MSBP has now banked practically all the UK's flora, for posterity.

Safeguarded species: The Triangular Club Rush (*Schoenoplectus triqueter*) once occurred in several tidal estuaries in the UK but is nearing extinction. Plants have been grown from seeds banked at the MSBP for reintroduction to suitable sites along the River Tamar. This action will improve its chances of long-term survival in the wild.

Bulgaria

The MSBP is working with the Institute of Botany, Bulgarian Academy of Sciences (IB-BAS). Between 2005 and 2009, 25 expeditions gathered more than 300 species. These have been dried, frozen and safely stored at the MSB, with some duplicated at IB-BAS. Bulgaria's flora is at risk from rapid development, as habitats make way for holiday resorts, ski tracks and golf courses.

Safeguarded species: The milkvetch *Astragalus physocalyx* is an endemic species on the Balkan peninsula and Anatolia and is listed in the Bulgarian Red Book. Its habitat (sclerophytic scrub) is threatened by livestock breeding and development. Seed collectors from the Institute of Botany, Bulgarian Academy of Sciences were able to collect seeds from this species in the Blagoevgrad district in June 2009.

China

The MSBP has helped China set up its largest wild plant conservation facility, the Germplasm Bank of Wild Species (GBWS). The Bank is managed by the Kunming Institute of Botany, of the Chinese Academy of Sciences, with the aim of conserving China's biodiversity and plant resources for sustainable use. The MSBP has given advice on construction of the new germplasm bank building, and is sharing its knowledge and experience of seed conservation. By 2010, 5,200 plant species from all over China had been banked, with 1,500 of them duplicated at the MSB.

Safeguarded species: The wild Yunnan banana *Musa itinerans* was the 24,200th species banked in the MSB vault. This number was a significant landmark because it meant that 10 per cent of the world's wild plant species had been banked.

Australia

Australia contains 15 per cent of the world's plant species; however, more than a fifth are threatened with extinction. The MSBP has partnerships with each of the six States and the Northern Territory. MSBP staff are sharing knowledge with 14 institutions and government departments on seed collection, conservation and research. These MSBP partners have mobilised a national network of seed conservation practitioners, now being actively supported by the Commonwealth Government under the auspices of the Australian Seed Bank Partnership.

Safeguarded species: The robust leek lily *Bulbine crassa* was only named in 2006. Only a few plants were known to exist, on islands off the southern coast of Victoria. In 2007, seeds were gathered from plants on Neds Reef in the Furneaux Islands as part of the Seed Safe project, a collaboration between Tasmania and the MSBP.

Jordan

MSBP scientists are complementing existing *in situ* conservation efforts by working with local partners to collect seeds, bank them and set up a mutually beneficial conservation training, research and educational programme. Kew's Geographic Information System team produced a Collection Guide to rare and endangered species of Jordan to help make finding sought-after plants easier.

Safeguarded species: MSBP scientists collecting seeds in Jordan became aware of the black iris *Iris nigricans* when a coach of tourists stopped to photograph it. The national flower of Jordan, with striking black flowers and bright green leaves, it is both endangered and endemic. Seeds are now safely stored at the MSBP and Jordan's National Centre for Agricultural Research and Extension.

The band of botanists who forge partnerships

A suite of collaborative projects, undertaken with over 100 partners in 50 countries, underpins the MSBP's work. Partner organisations range from the Lebanese Agricultural Research Institute and the Tbilisi Botanical Garden and Institute of Botany in Georgia, to the Chinese Academy of Sciences. When forging new relationships, the MSBP is motivated by the urgent need to bank seeds from the most useful and threatened plants, and enable use of these plants by researchers and communities. While working together with the MSBP, host nations receive help to develop their own seed-banking facilities and gain access to Kew's botanical resources and expertise.

The task of seeking out and forging new relationships lies with the MSBP's seven International Coordinators. This band of roaming agents travel the globe, identifying organisations that would make suitable partners, negotiating Access and Benefit Sharing Agreements (ABSAs) and, once a deal has been finalised, delivering the terms of the contracts. The process of developing each formal agreement can take anything from a few weeks to more than a year, depending on how complex the negotiations are.

'Some partnerships come about because the host countries have a particularly rich and interesting flora, such as Madagascar, USA, South Africa and Australia,' says International Coordinator Michiel van Slageren. 'We forged links with these nations because they contain large numbers of species and have great potential to help the MSBP meet its global seed-banking target. 'Partnerships in other countries might develop because an International Coordinator has an existing contact there. Sometimes we have had to halt a negotiation; this is often because the country in question has no framework in place to regulate the export of genetic material.'

Kew honours the letter and spirit of the Convention on Biological Diversity, the treaty adopted in 1992 at the first international Earth Summit in Brazil. It aims to conserve biodiversity, promote the sustainable use of plants and animals, and ensure that the benefits arising from genetic resources are shared out equally. These three principles guide how the MSBP develops its ABSAs with different nations. The overarching aim is conservation. In all cases, Kew does not use any genetic resources transferred to it for commercial purposes and it urges that recipients of samples from its collections follow the same principle.

The finer details of each ABSA vary according to each nation's needs. When Kew partnered with Namibia, for example, the country already had a well-developed seed conservation facility. Kew directed its efforts at helping train staff and was also able to offer expertise in identifying and classifying newly discovered species.

The situation with Mali was a different story. 'Everything that's there now, which includes a seed bank and herbarium plus three full-time staff, has all been supported by the MSBP,' says Michiel.

The MSBP's International Coordinators are expected to work towards delivering challenging plant-conservation targets. During negotiations with a potential new partner, they must try and identify the number of species the relationship is likely to safeguard and set that as a target. 'It must be realistic, not idealistic,' says Michiel. A balance must also be struck between the MSBP's global seed banking targets and the priorities of local partners.

ABOVE: Examining fruits in a head of *Tripteris microcarpa* in Namibia.

During one expedition to Egypt, Michiel was asked
to collect seeds from *Juniperus phoenicea*. Having
already gathered seeds from the species in Tunisia,
Lebanon and Saudi Arabia, he did not feel it was a
priority. However, the local collectors pointed out that
there was only a single population left in Egypt. 'It
would only have taken someone collecting wood for a
Bedouin pow-wow to wipe out the plant from Egypt's
flora,' he explains. 'It's at a moment like that you realise
you are part of a joint operation and that the wider
conservation context is also important.'

CONSERVATION FACT

The flowering inferno

Fires that raged across south-east Australia in February
2009 destroyed the only known population of the shrub
Nematolepis wilsonii, from a mountain forest 100 km east
of Melbourne. Fortunately, seeds of this species had
been collected in 2007 and banked at the Victorian
Conservation Seedbank (VCS) of the Royal Botanic
Gardens Melbourne, as well as at the MSBP. In all, 18,000
seeds were stored. An MSBP partner, the VCS is now using
the seeds to re-establish the plant. It has grown more than
150 specimens from seeds and cuttings, and planted them
close to the site of the original population. These will be
moved to the original site should no seedlings naturally
appear. It is unlikely seeds will have survived the fire as
the soil was burnt to a depth of one metre in places.

Tim Pearce, International Coordinator

East Africa and Australia

Tim trained as a plant ecologist in Aberdeen before working for the Wildlife Trusts in the UK. While at Aberdeen he did an ecological survey of the Afro-alpine mountains in Kenya and always hankered to go back. He was offered a post with the UK Department For International Development (DFID) to help the East African Herbarium in Nairobi to develop a plant conservation programme. That job started in 1992 and finished in 2000 but the Conservation Programme continues to evolve and he now manages the seed conservation programme in Kenya. The MSBP's relationship with Kenya really dates back to Tim's early days in Nairobi. Tim took a counterpart student, Patrick Muthoka, to the UK, who ended up studying for a PhD at Kew. MSBP representatives then came to Kenya to start partnership negotiations. When Tim's contract ended, the project was essentially taken under the wing of the MSBP and Tim was offered a role.

Through this partnership, five conservation and botanical institutes at the heart of Kenya's Seeds For Life programme are now collecting their native flora, banking the seed and initiating restoration schemes to help local communities and safeguard their glorious wild plants.

Tim and staff at the National Museums of Kenya, have carried out seed-collecting expeditions with the focus on iconic habitats such as Mount Kenya, the Masai Mara and the Rift Valley, and the endangered plants that live at the very tops of mountains: wonderful species of paper flowers, giant lobelias and giant groundsels, together with many endemic species. Working with the East African Herbarium, Tim and his colleagues have been building databases and helping to train Kenyan field collectors and botanists.

In Tanzania, species such as lobelias and African violets are part of the local flora. Today, there's a multi-million dollar industry throughout Europe trading in these plants, but in their native habitats they are highly threatened. The commercialisation of native flora is a key issue and cause for concern. With African succulents like the kalanchoes easily available in British supermarkets, it's easy to forget that these plants have a home far away. 'We are working with their custodians,' says Tim, who wishes the African conservation projects could get just one penny for every plant sold in the west.

In addition to his work in East Africa, Tim Pearce also coordinates Kew's work with Australian partners. Australia is a world leader in habitat restoration and its conservationists work to rehabilitate degraded bushland and safeguard native species, not least in the impressive King's Park restoration in Perth. And on the other side of the country, as part of SeedQuest NSW, the partnership has collected seeds from the Capertee Valley, a spectacular canyon at the western edge of the Wollemi National Park. The New South Wales seed bank collection provided seed from 20 species, which have been propagated as part of a national scheme to boost numbers of the Regent honeyeater. The restoration scheme is planting species of trees that are important sources of food for the honeyeater.

'As network coordinators, we at Kew can play our part in helping our partners distribute and communicate their knowledge. We have a great understanding of what's happening all around the world – a sort of helicopter vision – and we play a role in brokering experience of restoration across the network, getting data flowing from here to over there,' says Tim. 'We're a global botanic garden that just happens to physically be in south-west London and Sussex.'

Michael Way, International Coordinator

The Americas

In Chile, as elsewhere, housing and industrial developments are covering more and more of the unspoilt land. Today half of the country's native wild plants are either endangered or threatened, especially the geophytes (perennials that propagate via underground bulbs) and members of the daffodil family, the Amaryllidaceae, including spectacular rarities such as *Placea lutea*.

Michael Way is the MSBP's International Coordinator for the Americas. Fluent in Portuguese and Spanish, he's a veteran of many Kew expeditions to Chile, where he's helped to collect seeds from threatened and endangered native species, in partnership with a team of highly skilled Chilean botanists.

The current goal of the joint project is to conserve seeds from 20 per cent of the endemic species – those found nowhere else on Earth – and 60 per cent of the threatened species from Chile's desert and Mediterranean ecosystems. Southern Chile has great stretches of temperate forests while central Chile enjoys a Mediterranean-style climate that fosters a rich flora, which qualifies it as one of the world's biodiversity hot spots. The country also has desert lands, where rare species such as the cactus *Echinopsis chiloensis* are targets for the seed collectors. More than half of the plants of the coastal deserts in the north are found only in Chile.

Such botanical riches are packed into a small country with relatively few botanical specialists. 'The highly knowledgeable team in Chile are carrying out work of enormous dedication. They're working in a very systematic way, and provide the joint collection expeditions with all the botanical data we collectors need, including maps highlighting potential find spots,' Michael

says. 'The Chilean workers have amassed large seed samples from very rare and threatened species, returning to the same spots two or three times on occasion to coincide with the flowering and fruiting of the target plants. The intensity of their focus over several years is remarkable.' Seed samples are collected, processed and banked in the seed bank in the National Agricultural Research Institute (INIA), with duplicates sent for banking in the MSB. So far, seeds from 740 native species from Chile have been added to the stores at Wakehurst.

Michael also coordinates the MSBP's links to Mexico, where arid and semi-arid lands are home to some 6,000 native plant species; approximately 20 per cent of the biodiversity of Mexico. 'The human pressures on the Mexican drylands are intense, and it is urgent that we continue to expand our project activities in Mexico and strengthen our links with other organisations and with local communities,' he says, an aim in part achieved through the Useful Plants Project (see page 152–157). Meanwhile, seed collectors are focusing their efforts, thanks to work by Oswaldo Tellez, who has worked with Michael and the MSBP team to prepare target guides. So far, three Collection Guides have been completed covering the Sierras of Taxco and Huautla, Sierra Gorda-Rio Moctezuma, and the Peninsula of Baja California.

In the USA, Michael links up with a complex web of organisations, many of them a part of the Seeds of Success programme. Established by Kew and the US government's Bureau of Land Management (BLM) in 2001, the Seeds of Success programme coordinates seed collections of native plant populations in the USA for conservation and habitat restoration. One project aims to restore 400 million acres of public land damaged by fire, overgrazing and industrial development. There are

ABOVE: **Recording flowers of** *Mutisia* **sp. with a data sheet.**
RIGHT: **Michael admiring the rare** *Plazia cheiranthifolia* **in Chile's mountains.**

huge challenges, and a great need for native plant material for re-establishing original plant communities.

The MSBP has helped to train an army of botanists, both professional and amateur, and advises them on when on to collect and which species to send to the MSB as globally new additions. 'All the MSB seed and herbarium material collected by BLM comes first to Wakehurst by Fed Ex,' explains Michael. 'Once received, we compile the data, clean the seeds and send half of each collection back for use in the United States Department of Agriculture germplasm system.'

Kew also works with nongovernmental partners and volunteers on more than 100 registered collections at 60 sites, all managed to some extent from the MSB, and trained using a programme based on a curriculum devised by Kew. 'Right across the USA – from Washington DC to Chicago, from the Lady Bird Johnson Wildflower Center in Texas to San Diego and Phoenix, from New England and New York through the salt marshes on urban fringes, and prairie, desert and wildflower meadows – the power of volunteers is the story,' says Michael.

CONSERVATION FACT

Conservation in the USA

In the Midwestern USA, the tallgrass prairie is now exceedingly rare, with only one tenth of one per cent of the original habitat remaining. Conservationists in Chicago are collecting seeds from 1,500 prairie species in a bid to recreate the dramatic plant community that once supported vast herds of buffalo. In the USA overall, one in five native plant species are threatened with extinction.

Michiel van Slageren, International Coordinator

West and Southern Africa, Australia and the Middle East

Michiel van Slageren juggles MSBP projects in Jordan, Lebanon, Namibia, South Africa, Tasmania and the Northern Territory of Australia. It's a role that requires him to travel for around a quarter of each year, with the remainder of his time spent at Wakehurst Place 'under the pressure of email and administration'.

I like to show up once a year to discuss with each partner how we can further the project, and to undertake a joint collecting trip,' he says. 'I feel it's important that I go out in the field and join in with everyone else, rather than sitting in my comfortable hotel room handing out orders then going back home.'

Michiel, who describes himself as a 'Dutchman of advanced age' undertook an MSc and PhD at the University of Utrecht in the Netherlands before working for six-and-a-half years at the International Centre for Agricultural Research in the Dry Areas (ICARDA) in Aleppo, Syria. Here, he collected wild relatives of wheat on behalf of the Dutch Government. In 1995, he applied for the role of Seed Collector for Africa at Kew and got the job. 'In this role I saw the seed bank project slowly developing and remember the day Roger Smith announced it had been awarded funding by the Millennium Commission,' Michiel recalls. 'I later applied for the role of International Coordinator for the MSBP.'

Michiel negotiated the first Access and Benefit-Sharing Agreement (ABSA) for the MSBP with Egypt in 2000 and subsequently developed collaborations with partner organisations in Lebanon, Jordan, Saudi Arabia, Burkina Faso, Mali, Australia's Northern Territory and Tasmania. His present role involves negotiating new mutually beneficial agreements between Kew and partners, developing procedures for 'best practice' during joint seed-collecting missions, training staff in

partnership organisations, meeting plant conservation targets, and helping develop *ex situ* seed-banking facilities. 'Each International Coordinator brings different qualities,' he says. 'In that sense I feel I am one of Kew's Arab world experts.'

The projects Michiel works on have varying goals. The MSBP's ten year collaboration with Jordan's National Center for Agricultural Research and Extension (NCARE) aims to complement existing *in-situ* conservation efforts by setting up a seed bank of indigenous wild plant species and creating a conservation training, research and educational programme. Experts at Kew used digital maps to produce a Collection Guide to Jordan's rare and endangered species; this has helped the MSBP collect and safeguard some 35 new species from Jordan each year. In Namibia, one of the goals of the MSBP's partnership with the National Botanical Research Institute is to develop a conservation strategy for the important succulent species *Salsola nollothensis* ahead of planned mining developments.

In total, Michiel has undertaken 70 expeditions since 1995, including multiple visits to project partners. He enjoys the diversity of the places he visits and revels in the different people and cultures he encounters. He is well aware that an expedition's success is often thanks to the valuable help that partners and other locals contribute, often at short notice. On one occasion in Burkina Faso, the team had collected so many seeds there was not enough room in their vehicle to carry them all. Members from the local partnership quickly organised to send the material by train from the country's second city Bobo Dioulasso to the capital Ouagadougou. 'By the time we got back people were already working to clean up the seeds,' says Michiel.

PLANT STORY
Dangerous collections

Michiel's seed-collecting journey to Lebanon in 2006 coincided with the outbreak of war between Israel and Hezbollah. The Israelis bombed the runway at Beirut airport and closed all the borders with Syria, making return to the UK somewhat difficult. Michiel and his colleague Tiziana Ulian were in the centre of the country when the hostilities started. In the end they were evacuated by the Italians.

'We left in a convoy of buses that were allowed out through a border crossing into Syria,' Michiel recalls. 'The Israelis considered every truck as a vehicle transporting grenades and rocket launchers around, which did rather limit our enjoyment of travelling on Lebanon's roads. However, there was apparently a deal made between the Israelis and the Lebanese that they would not bomb the convoy on our journey to Lebanon. After our departure the Israelis bombed the border post we had used.

'We then flew by Italian military transport plane to Cyprus' says Michiel. From there we travelled to Rome where, as officially recognized refugees, we were given a reduced-price ticket to London. Looking back, we were lucky to have been outside the combat zone; we were not bombed and no one was hurt. However, since then Kew has sharpened up its advice for staff who visit places that could be dangerous. Some people think I'm always seeking danger and excitement but it's only in the sense of having a good anecdote to tell people later.'

It's not unusual for local partners in Africa to hire the best part of a whole village to clean up seeds bound for the UK. On an expedition in Burkina Faso that Michiel participated in, the team came across a 20-metre-high palm tree that had huge inflorescences with large seeds. 'After getting permission to take some fruits we had to find a specialised climber who could obtain them for us. Eventually we had so much material we had to hire people with machetes to cut off the fruit from around the 200 or so seeds so it wouldn't cost us a fortune to ship them back to the UK. I still use those seeds at the MSBP to explain the difference between orthodox and recalcitrant seeds or to highlight the importance of palms. It's a great privilege to know the entire story of how those seeds came into our hands.'

OPPOSITE: **Gotcha! Michiel collecting** *Microcachrys tetragona* **in Tasmania, Australia.**

ABOVE LEFT: **checking the quality of seeds in Namibia.**

ABOVE: **Trying to avoid capsizing on lakes near the Niger river in Mali.**

Moctar Sacande, International Coordinator

Africa, including Madagascar

Making excellent use of his varied expertise, Moctar Sacande, the International Coordinator for most of Africa including Madagascar, has been investigating ways of improving the storage of tropical tree seeds. Born in Burkina Faso, and fluent in French and several West African languages as well as English, Moctar is a seed biologist as well as an expert in tropical forestry.

One of the biggest technical barriers to sustainable forestry is the difficulty of storing tropical seed from year to year. Tropical trees tend to flower and set seed irregularly, typically every three to seven years. Unfortunately, they are often resistant to drying (recalcitrant) and their storage life tends to be measured in months at best. Maturity can be an important factor: Moctar uses neem (*Azadirachta indica*) as a model species and has discovered that neem seeds taken from yellow fruits, rather than green and brown ones, survive significantly longer in storage.

Moctar has been helping to coordinate research into the conservation and sustainable use of native tree seeds of great value to local communities, targeting 60 priority tree species across 16 sub-Saharan African countries. The project, part of the Darwin Initiative, is promoting the propagation of indigenous species that are better adapted to local conditions and much more useful than exotic, fast growing species which are often increasingly rare, such as the shea butter tree *Vitellaria paradoxa*. The researchers are discovering which species can be dried and still germinate (orthodox), allowing long-term storage of their seeds, and devising ways to maximise the storability of recalcitrant species.

Moctar has also fostered the spread of the MSBP throughout French-speaking West Africa, through workshops that include databasing, seed banking and habitat restoration for scientists and technicians from countries such as Mali, Burkina Faso and Madagascar. Resulting collaborative projects and field collecting trips have included one in which two teams of collectors, each six members strong, travelled over 1,600 kilometres each, northwards and southwards, starting along the River Niger in Mali and reuniting in Ouagadougou, Burkina Faso. A record collection of 100 wild species was made, including important and threatened ones in the Sahel. Large proportions of these countries' unique floras are now conserved, in their native lands and at the MSB.

The progress made in Mali in recent years is particularly striking given that, until recently, the country didn't have a seed bank. So far, conservationists have collected 500 species, 20–25 per cent of flora. Moctar travels out to the partner countries two or three times a year, helping with decision-making regarding where to target next, plus collecting and training.

Madagascar is one of the MSBP's key partnerships. The island has a stunning unique flora. Eighty per cent of its species are endemic; with 10–12,000 species, there is still a lot of work to do there. So far, the teams have collected more than 1,000 species, approximately 10 per cent of flora. 'We may yet find completely new species,' says Moctar. The focus is on especially rare and endangered plants in various families, including that iconic tree, the baobab (*Adansonia digitata*). 'There are eight species in the genus, and we've got them all,' he says.

Meanwhile in Malawi, which has lots of endemics in its mountains, the field collectors have collected over 600 species. Malawi also recently hosted a regional meeting of seed bank partners, followed by an expedition to Mount Mulanje, at 3,000 metres, the highest massif in south-central Africa.

ABOVE: Tree ferns in Malawi.
OPPOSITE: Moctar planting Mongongo (*Schinziophyton rautanenii*) trees in the Kalahari desert of Botswana.

Clare Trivedi, International Coordinator

Other countries

BELOW: **Clare collecting seeds in Kenya.**

OPPOSITE: **A stormy future:** contemplating the impact that climate change will have on plant life in the coming years.

In recent years many new countries have joined the MSBP, including Slovakia, Bulgaria, Georgia, Kyrgyzstan and Canada. All these countries are fully signed-up partners to the project, but with agreements renewed year by year, rather than on the longer time scale of the countries that joined at the project's inception. In her role as International Coordinator, Clare Trivedi has been building capacity in these 'other countries'. Their contribution has been vital. The MSBP achieved its first landmark of collecting seeds of ten per cent of the world's flora (24,000 species) in 2010, thanks in no small part to the seeds collected by these newcomers in recent years.

Clare and her fellow International Coordinators act as conduits linking local botanists, seed technologists, farmers and horticulturalists with experts at the MSBP.

During joint seed-collecting missions and on-the-job training of Kew's counterparts abroad, Clare helped to foster best practice on more than 22 expeditions between 2001 and 2005. Now she tries to ensure she re-visits each country at least once a year.

The advance made in seed conservation in some countries is impressive, says Clare, citing Georgia as an example. 'Georgia has amazing botanists, they know their flora inside out, but the seed collection at the Tbilisi Botanical Garden and Institute of Botany used to be small and poorly documented, and the technology for drying and storing seeds less than ideal.' Now their work is in line with international standards and Kew's input has helped raise the profile of Georgia's conservation project. When Clare first visited, seed collecting was not seen as a priority, and the status of seed collecting was low. That has now changed and the work is better paid.

Meanwhile, Georgian botanists are working to improve their germination and propagation facilities, so that populations of alpine rarities threatened by climate change can ultimately be replenished in the wild. 'It's the whole cycle, that is the rewarding part of the job, seeing progress and making friends,' says Clare.

It's hands-on conservation work, which is exactly what inspired Clare when, as a teenager, she first heard about the ozone-layer hole and rainforest destruction, and decided to do what she could to tackle the world's ecological nightmares. She studied environmental sciences at university then worked on climate-change policy for the Royal Society and an NGO. Now she's at the forefront of Kew's response to the coming crisis.

The message we need to give out, Clare explains,

'The message we need to give out is that climate change is going to have a big impact on biodiversity. It's expected to be the biggest driver of biodiversity loss over the next century.'

is that climate change is going to have a big impact on biodiversity. 'It's expected to be the biggest driver of biodiversity loss over the next century.' she warns. Yet predicting what's going to happen to any individual species can be difficult. 'That's where the seed bank and *ex situ* conservation gives us a safeguard in an uncertain world and provides options for the future.'

With so many of the world's wild plants under threat, the need to recruit more of the world's nations to the conservation cause is intensifying. Already, Clare has helped to build bridges between the team at Wakehurst and Chinese seed technologists and botanists. She's been part of the MSBP team providing advice and support to China's largest seed bank: the Germplasm Bank of Wild Species at the Kunming Institute of Botany, which was launched in 2008.

On the MSBP wish list for future collaborations, India and Brazil loom large. So far, neither country has teamed up with the MSBP to collect seeds, out of a reluctance to allow duplicate seed collections to come to the UK. But in the second phase of the MSBP, which started in 2010 and will run until 2020, countries will be able to opt out of that requirement. 'As a result, the project will become a true global network,' says Clare. 'What's important is that all the collecting and conserving is done to the highest international standards, not that all of it is in the building at Wakehurst.' The aim is to move to a self-sustaining network that will be truly international.

What's more, the current, second phase, will focus not only on making collections, but on using them in restoration and sustainable use projects all over the world. 'Our mission is conservation, we don't just want to stick seeds in the bank, we want to see them out there, helping people and restoring ecosystems,' Clare says.

SYRIAN BEAR'S BREECHES

LEBANON, JULY 1998: Michiel van Slageren, of the MSBP, and Simon Khairallah of the Lebanese Agricultural Research Institute (LARI), are driving along the main road near Jubb-Jannine in the southern part of the Beka'a Valley. They are on a seed-collecting expedition, but one plant eludes them: *Acanthus syriacus*. So far, they haven't seen a single plant, let alone seed. But as they drive along, they come across six plants beside the road. Has their luck turned? Not this time, as, to their great disappointment, none of the plants spotted bear any seed.

The hunt for Syrian bear's breeches seed continues for four years. Teams from the MSBP and LARI organise many trips to various Lebanese provinces to collect the seed of wild *Acanthus syriacus*, but find no suitable plants. They even return to the six plants by the road near Jubb-Jannine, but all those plants have been destroyed. Collecting the seed of *Acanthus syriacus* is turning out to be quite a challenge.

In the spring of 2002, Simon Khairallah and Joêlle Breidi from LARI, on a collecting trip in south-eastern Lebanon, find plants growing at three sites. The best site has about 50 plants on the border of a wheat field, so they keep an eye on them, returning to the site several times while they wait for seed to develop. Everything seems fine, but just as they are about to collect seed they discover that all the plants are either diseased or damaged by insects. The same is true for the second site.

THIS PAGE: Simon Khairallah and, below, views of the Beka'a Valley.

Overcoming their disappointment, the team from LARI travel to the third site. Here they find plants that have not been cut, ploughed, diseased or damaged by insects but that are intact, healthy and yield enough seed to make a decent collection.

Thanks to the seed collectors' determination and persistence over the four-year search, *Acanthus syriacus* seeds are now safely stored in seed banks.

THIS PAGE: The elusive plant is finally in the bag.

Plant profile

COMMON NAME: Syrian bear's breeches, Syrian acanthus

LATIN NAME: Acanthus syriacus

FAMILY: Acanthaceae

STATUS: Not currently assessed by the IUCN, but considered by MSBP staff to be endangered in Lebanon

THREATS: Agriculture and loss of habitat

SIZE: 20–80 cm tall

DESCRIPTION: Acanthus syriacus is a spiny perennial herb with large purple and white flowers in a spectacular spike up to 60 cm long. The leaves are clustered towards the base of the stems, often forming a loose rosette. The lance-shaped leaf is leathery with a hairy surface and about 10–30 cm long. The leaves are deeply lobed, each lobe being toothed and tipped with a harsh spine. The fruit is a four-seeded capsule.

Local botanists build a global seed bank

In its early days, the MSBP employed seed collectors. With its initial target to bank seeds from all of the UK's wild flora, it despatched these field botanists to Britain's meadows, woodlands, rivers and lakes to gather seeds of everything from sneezewort (*Achillea ptarmica*) and chaffweed (*Anagallis minima*) to horned pondweed (*Zannichellia palustris*) and dwarf eelgrass (*Zostera noltii*). Today, however, with most of the UK flora stored for posterity, the MSBP's work abroad has taken centre stage. So successful have its International Coordinators been in forging partnerships, negotiating Access and Benefit Sharing Rights and training staff in partner institutions around the world that they no longer need to be so hands on. The majority of seed collecting is now carried out by collectors employed by and stationed with local partner organisations.

'We all still do some seed collecting to keep that intellectual capital,' explains former seed collector and International Coordinator Tim Pearce. 'But now I see fantastic seed collecting being carried out by our partners all over the world. That's important as it's the end product of all the hard work we've put in in terms of training. Some of the guys abroad are highly experienced world-class seed collectors, while others are just starting out on the ladder.

The seed collecting process is a fantastic way to learn your trade. In Kenya, where my experience lies, I've watched people come out of school with the equivalent of their GCSEs and diplomas and join the seed collection teams through the local herbarium or forestry department. They are required to really understand the workings of herbaria, grasp basic taxonomy and seed biology and to hone their field identification skills. The whole programme has really helped those young people progress in their careers.'

MSBP partners typically comprise three types of organisations. There are herbaria, who are interested in wild plants and biodiversity; traditional gene banks, which tend to have an agricultural focus and staff who are used to working with crops such as rice, millet, corn and pigeon peas but who are beginning to investigate wild crop species; and forestry seed centres, which supply farmers, plus private or government forestry trades with trees and are involved with the timber industry. 'The MSBP typically straddles all three as it requires people who understand about wild species, have experience of seed banking and have knowledge of how plants are used,' says Tim. 'Generally an expedition will comprise a field botanist from the herbarium, someone from a forestry centre and some assistants. In Africa, occasionally there might be an armed guard if the site is likely to be inhabited by elephants or buffalo.'

Before any seeds can be collected from the wild, collectors need to obtain permission. In many countries, particularly if collecting in protected areas or those rich in biodiversity, an official seed-collecting permit is a legal requirement. All mandatory permits are obtained before heading out on an expedition but once the team arrives at a site, if it's privately owned, they need to get permission from the landowner.

ABOVE: **MSBP staff collecting** *Potamogeton polygonifolius* **in the Upper Teesdale, UK.**
ABOVE RIGHT: **Patrick Muthoka and Mathias Mbale collecting in the species-rich Tiata Taveta region of Kenya.**

Usually this involves the team leader going to explain that the team is on official government business. Where possible, they try to give something in exchange for access. 'We frequently employ the children of the landowner,' explains Tim. 'This has two benefits as the landowner receives money and the children are taught something. That's really the notion of benefit sharing. We try to bring everyone on board.'

One of Tim's favourite seed-collecting recollections relates to the *Kalanchoe*. This genus contains around 125 tropical flowering succulent species, some of which are sold in the UK as houseplants. In the wild, they grow primarily in southern and eastern Africa and Madagascar. One particular species, *Kalanchoe bipartita* had been collected once in Somalia and twice from the same location in Kenya but had not then been seen since the 1970s. 'I lived in Kenya for ten years and looked for it pretty much every season in the place it had been recorded. Never found it,' recalls Tim. 'Then low and behold guys from our Kenyan partners turned up a fantastic population at another site. It was great to find the plant but I was even more pleased that these boys that I'd watched growing up from school leavers took the initiative to go and look for it. Thanks to them, we've now got a fantastic seed collection and the conservation of the species is now assured.'

Narrowing the search

When partner countries need help targeting species for collection, the MSBP's Species Targeting Team (STT) steps into action. The team uses herbarium specimens from partner herbaria and the significant collections at Kew's Herbarium, along with various literature resources, to gather information needed to help prioritise species. In its first three years, the STT produced Collection Guides for Botswana, Burkina Faso, Chile, Jordan, Kenya, Lebanon, Madagascar, Malawi, Mali, Mexico, South Africa, and Tanzania. As part of the process, the scientists produce conservation assessments for targeted species so they can be prioritised for collection according to their risk of extinction. For each country, those species that are most likely to become extinct undergo a fuller conservation assessment based on IUCN Red List categories and criteria. These are submitted to the IUCN Species Survival Commission for review and sent to partner institutions for use in conservation management programmes.

BELOW: Patrick Muthoka and Mathias Mbale collecting in the species-rich Tiata Taveta region of Kenya.

Dan Duval

Seed Collector

Botanic Gardens of Adelaide

A Seed Collector at the Botanic Gardens of Adelaide (BGA), Dan Duval has been involved with the MSBP since 2004. The BGA had become a partner to the MSBP the previous year, a relationship formalised as the South Australian Centre for Rare and Endangered (SACRED) Seeds project. The partnership enabled BGA to fund a full-time position, and Dan's experience in seed germination techniques made him a suitable candidate. 'The timing was perfect as BGA's seed collection centre started just as the MSBP were looking for overseas partners,' he says.

Funds provided by the MSBP and South Australia's State Government enabled BGA to equip laboratories for drying seeds, testing their viability and germinability, and storing them at low temperatures.

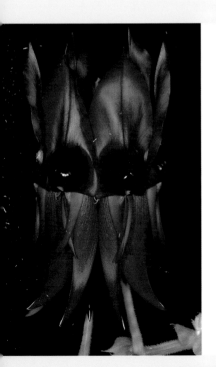

AMAZING PLANT FACT

Desert beauty

South Australia's state flower is Sturt's desert pea, *Swainsona formosa*. The genus is named after the English botanist Isaac Swainson, who became famous for his botanical garden at Twickenham, London. It was funded largely from sales of a medicine called Velnos's Vegetable Syrup, which was allegedly able to cure everything from leprosy to tape worms. The plant's common name derives from Charles Sturt, who observed large numbers of Sturt peas while exploring Australia in 1844. Seeds from the plant are now secure in seed banks in Adelaide, Perth and at the MSBP. *Formosa* is Latin for beautiful; the plant has vibrant red flowers that stand on stalks some 30 cm above the desert sands.

Dan's role involves juggling the tasks of collecting seeds, often from remote locations, when weather and environmental conditions dictate, and processing the material brought back for inclusion in the BGA's Herbarium and seed bank. If seeds are available in sufficient quantities they are also sent to the MSBP, where they contribute to its seed-collecting targets and provide a back-up collection.

Field collecting starts in late winter or early spring (August or September) and continues through to February. Dan and his colleagues initially set itineraries, second-guessing nature to time their arrival at target habitats when weather conditions are optimal for plants setting seeds. At this time of year they travel every second week to locations up to 2,000 km away, filling the time in between with visits to closer sites. Although itineraries are planned in advance, the collectors have to be prepared to react quickly if unexpected opportunities arise.

'We're going away for an unplanned overnight trip next weekend because there's been some rain in the Flinders' Ranges and we have the opportunity to relocate the rare creeping darling pea (*Swainsona viridis*), which we've been after for a long time,' explains Dan. 'It's the first reasonable rainfall in this region for ten years. We're going to drive overnight, camp and travel back again. It's a long way from here but it will be a cost-effective trip because we won't pay for accommodation, we'll just swag it. The plant is endemic to South Australia and is listed as vulnerable. It occurs throughout eastern and central parts of the Flinders' Ranges but we've only seen it once before.'

The SACRED seeds project has a good track record of finding and conserving rare species. An example is

'It's very rewarding to think that, in my lifetime, I've done something to prevent a species from going extinct.'

that of the spade-leaf bittercress, *Cardamine gunnii*. This endemic species was considered to be extinct in Australia, although it had been historically recorded in Victoria and South Australia. Although there appeared to be a few recent records, these were verified as being other closely related species. However, a single 1980s record from a redgum swamp in the south-east of the state was still in question.

BGA wanted to revisit the site and check if the plant in question really was *Cardamine gunnii* or a similar-looking species that had simply been mistaken for it. At the time, the species was listed as vulnerable but there were no records to substantiate that rating. Dan and his colleagues went to the spot, confirmed that the plant was indeed *Cardamine gunnii*, and have since nominated the species under the Environment Protection and

Biodiversity Conservation (EPBC) Act as endangered.

'As a result of our work, there will now be some ongoing conservation work at that site,' says Dan. Once a plant is listed it gets protected federally, so that means there's a lot more legislation in place to conserve it. People who work on the threatened forests that contain *Cardamine gunnii* will be able to get funding to search for more populations and there will be fact sheets created to raise awareness.

What I like about my work is that I get the chance to find species first-hand in the field, learn something about them, work out how to germinate them and then begin propagating them to produce plants. We're now putting some of the rare plants we've found back into project sites. It's very rewarding to think that, in my lifetime, I've done something to prevent a species from going extinct.'

How botanists gather seeds from the world's precious flora

BELOW: *Datura quercifolia,* the oak-leaf thorn-apple.
OPPOSITE: Andrew Crawford, collecting seeds in Western Australia.

With as many as 100,000 types of plant facing extinction, the MSBP's aim of collecting 25 per cent of the world's flora by 2020 is a race against time. Every year, some 200 seed-collecting expeditions are staged by the MSBP and its partners to try and obtain seeds from the world's most rare, threatened or useful wild plants. Efforts to date have resulted in seeds from 135 countries, ranging from Namibia and Pakistan to Lebanon and the Falkland Islands, being stored at the seed bank. Often to remote, sometimes dangerous, places, seed-gathering expeditions take considerable planning to ensure success.

Before heading into the field, collectors must gather information about the plants they wish to target. The more they know about the favoured habitat of the plant, what it looks like, the season in which it fruits and any environmental conditions likely to induce seed production, the more successful the expedition is likely to be. 'Here in Australia, many plants are responsive to

fire,' explains Dan Duval, a Seed Collector at MSBP partner the Botanic Gardens of Adelaide. 'It doesn't matter how hard we look for certain plants, we'll only see them if there's a fire. Suddenly, there'll be a once-in-50-years fire and we'll find a species we're after in numbers. Then, the next year it'll be gone again.'

Once collectors encounter a plant they think they need seeds from, they have to correctly identify it. They consult checklists and field guides and make detailed descriptions of the plants while in the field. They also take a sample of the plant and press it for inclusion in Kew's and partners' herbaria. In addition, photographs, seeds or DNA are sometimes taken. This dried plant sample, called a voucher, is what taxonomists back at Kew's Herbarium or the MSBP use to confirm that the name given to the seed collected in the field is correct.

Having identified and named a plant, the collectors must decide whether there are sufficient seeds of good enough quality to take a collection. This involves cutting some seeds open to estimate the percentage that are damaged, infested or empty. If a collection gets the go-ahead, the scientists take seeds from a random selection of at least 50 plants. Ideally, they try to take 20,000 or so seeds, so that both partners and the MSBP have sufficient quantities to send samples out for research. However, they aim to take fewer than 20 per cent of available seeds in the population so as not to jeopardise any plant's future survival in the wild. 'It sounds a lot but 20,000 seeds is just a spoonful for some species,' says Dan.

The type of plant and how it naturally disperses its seeds generally dictates how collectors gather seeds for the MSBP. They might use long-handled pruners to snip seeds clustered in high tree branches, pluck fleshy

AMAZING SEED FACT

Little and large

The largest seeds stored in the MSBP are 60 mm across and come from a *Hyphaene* palm. Sometimes called the ilala palm or vegetable ivory palm, the species within the genus grow across Africa, the Middle East and India. The smallest seeds held in the bank are orchid seeds of less than 1 mm in diameter. Unlike the palm seeds, these are like specks of dust, and have to be counted and sorted under a microscope.

fruits individually by hand into buckets or shake wind-dispersed seeds into large tarpaulins for collection *en masse*. If the seeds are not quite ready for shedding, collectors sometimes tie cloth bags around seed heads and come back in a few days' time to retrieve the bounty. The coordinates of every collection made are recorded using a GPS (global positioning system), along with directions on how to find the plants again and information on dominant plant species at the locality, soil type, slope and land use.

Seeds must be carefully looked after in between being taken from the field and delivered into a local seed bank or the MSB. High temperatures and humidity can damage seeds, so collectors often lay collections out to dry in a shady, well-ventilated spot. Those within fleshy fruits are usually extracted within one or two days of collecting to reduce the risk of mould damaging them. If there are many seeds of this kind, it can be a time-consuming task. Most MSBP partners now have the capacity to process seeds; they do so, then send duplicate sets of seeds to the MSBP for storage. Only seeds gathered in the UK, plus a few from the USA, are sent directly to the MSBP for processing these days.

Tracking down a lost plant in China's limestone country

RIGHT: Fruit of *Illicium simonsii*, one of Yunnan's many sought-after species.

OPPOSITE, CLOCKWISE FROM LEFT: The Auspicious Light Pagoda in Suzhou, China; the MSBP's Kate Gold instructing Chinese partners on collection techniques; Chinese partners putting seed-gathering theory into practice.

In 1906, while working as a missionary in China, Frenchman E. Ducloux gathered a specimen of a limestone-loving plant with small yellow flowers. It eventually ended up among the botanical collections of the Muséum National d'Histoire Naturelle in Paris, identified as a member of the Gesneriaceae family. In 1997, Gesneriaceae expert Professor Wen-Tsai Wang, of the Institute of Botany in Beijing (part of the Chinese Academy of Sciences) examined the specimen while working on the *Flora of China*. He saw that its corolla lobes and stamens were arranged differently to other Gesneriads and realised it represented the sole member of a new genus. He published his finding in the botanical press, naming the plant as *Paraisometrum mileense*.

At this point, the plant was only known from that one specimen. Endemic to China, it was either extremely rare or had gone extinct since Ducloux had gathered it. There was only one way to find out if it still existed: go and look for it. In 2006, Dr Shui Yumin and students from the Kunming Institute of Botany set out to try and find living specimens of the species. Most Gesneriads only inhabit nutrition-poor and vulnerable limestone habitats in Yunnan, so this is where they looked. It was a wise decision; they finally tracked down the wild *P. mileense* in Shilin County, south-eastern Yunnan. The scientists counted 320 plants of the species at the location, of which around a third were in bloom. They collected seeds from some of the plants to try and save the species from extinction. Located close to settlements, *P. mileense* is at risk from human activities.

The scientists reported their find to local and national government representatives. In 2009, the State Forestry Administration of China launched the 'Conservation of Species With Extremely Small Populations' programme. *Paraisometrum mileense* was listed in the top 20 prioritised plant species under this conservation initiative. The same year, a new population was discovered in Guangxi province. The rediscovery in the wild of a species previously thought to exist only as a herbarium specimen provides botanists with an opportunity to try and understand the plant's evolution, conservation and potential uses. Ongoing studies have revealed that the plant has an unusual pollination regime and unique reproductive mechanism. Seeds will be stored in the Germplasm Bank of Wild Species (GBWS) in Kunming and the MSBP. Seedlings will also be grown and transplanted into Kunming Botanic Garden to raise awareness among visitors.

Joint effort to conserve rich biodiversity

China is the third most biodiverse country in the world, home to over 30,000 species. A ten-year agreement was signed between the MSBP and the GBWS in May 2004. Under this agreement, 4,000 threatened and endemic plant species from China have been collected and safeguarded in Chinese facilities, with half duplicated in Kew's MSB. The collaboration has encouraged UK and Chinese botanists to work together to collect and conserve China's plant diversity through a range of scientific activities including staff exchanges, PhD co-supervision, conservation skills training and information sharing. Scientists at the MSB provided technical support during construction of the GBWS building, and continue to share knowledge and experience of seed conservation activities.

Bringing the world's rare orchids back from the brink of extinction

The orchid family (Orchidaceae) is one of the largest and most diverse families of flowering plants. Orchids constitute in the region of 10 per cent of the world's flowering plants, roughly 25,000 species, which are unevenly distributed throughout the world. The UK has some 50 wild orchids, while Mount Kinabalu in Borneo has a whopping 850 species of orchids alone. Many beautiful species were so heavily collected by Victorian botanists that all that remains are faded paintings or a few pressed specimens in herbaria. Sadly, pressure on wild orchid populations is intensifying all over the world as key habitats are destroyed. Rare species continue to be illegally ripped from forests to supply the trade in traditional medicines and in horticulture. To make matters worse, many orchid habitats are in imminent peril because of global climate change.

No one can be certain precisely how many wild orchids are teetering on the edge of extinction but in Australia's state of Victoria alone, 208 out of 372 orchid species are threatened or already extinct, while on the slopes of Mount Kinabalu in Borneo, *Paphiopedilum rothschildianum* has been reduced to just three populations. Slipper orchids are particularly at risk from unscrupulous collectors, and the need for action is urgent.

In 2007, seed scientists and orchid enthusiasts around the globe began to work together to organise a global network for storing seeds of the world's orchids. The project, headed by Hugh Pritchard and Phil Seaton at Kew's Seed Conservation Department, is called Orchid Seed Stores for Sustainable Use (OSSSU). The project secured funding for three years through the Darwin Initiative project, part of the UK government's response to the Rio Earth Summit.

The project aims to establish a self-sustaining global network of orchid seed banks, ensuring a future for the world's irreplaceable orchid species using state-of-the-art techniques and facilities. Phil Seaton, an orchid enthusiast and former lecturer in biology, gives voice to the motivation behind the project, saying: 'I want my

The tiniest of seeds

Orchid seeds are tiny, rivalling the smallest dust particles. A collection of seeds weighing just one gram, and occupying just a few cubic centimetres, could grow into 350,000 individual plants. The entire world's orchid flora could easily be stored in just three chest freezers. Orchid seeds are minute because they have evolved to be carried away on the gentlest of breezes. As the seeds do not contain food reserves, germinating seeds access nutrients by teaming up with a fungus.

children and grandchildren to be able to experience what I have enjoyed. We don't have the right to deprive future generations of the ability to appreciate these wonderful plants.'

This relatively inexpensive insurance policy will ensure that orchid seeds are available in the future for reintroduction, habitat restoration and sustainable use in their countries of origin. As a desirable spin-off, they could also be used as a reservoir of material to encourage the raising of orchids from seed, thereby reducing the pressure on wild populations from unscrupulous collectors. 'There is no time to delay,' says Phil. 'We know that most orchid seeds can withstand drying much like conventional crop seeds, and so potentially can be stored in seed banks at refrigerator or freezer temperatures for many decades. What's more, many orchid species can be raised from seed with little more equipment than can be found in the average kitchen.'

The goal for the project's second phase, starting in autumn 2010, is to conserve 1,000 species of orchids.

This will be achieved by expanding the network to approximately 40 institutes, doubling the number of countries in the network from the current 15 to around 30. The network will form a Global Orchid Facility, promoting orchid conservation and serving as a scientific and educational resource. Partner institutions are already gathering information on the challenges of orchid seed germination, and research into *in vitro* germination of orchid seed, to aid germination through the provision of sucrose and micronutrients, is under way.

Commercial and amateur growers, as well as orchid specialists from universities and botanic gardens, will all play a vital role. 'There are a lot of enthusiastic people out there,' says Phil, 'people from all walks of life, and we need to include them all.' There is something special about orchid people, he reckons, with their particular dedication and enthusiasm. 'If we are to make a real impact and rescue our botanical riches for future generations, we need a great cooperative venture, uniting hobbyists, commercial growers and scientists.'

LEFT TO RIGHT: **More flagship orchid species:** *Encyclia phoenicea* from Cuba; *Lycaste skinneri* from Guatemala; and *Rhynchostylis gigantea* from Thailand.

Desert mission safeguards rarities

OPPOSITE: **The wide open road: Namibia's steppes in the central-west part of the country.**

The Sperrgebiet in Namibia's south-west corner is a fantastic place to collect seed. Not only is it 26,000 square kilometres of restricted area – no roads, no railways and its own security force to prevent illegal diamond mining and trafficking – it is also a biodiversity hotspot. About 776 plant species grow there, of which 234 are endemic and 284 are on the IUCN's Red List of Threatened Species. This is exceptional considering the area has only 100 mm of rain a year and gets most of its moisture from fogs that roll continually in from the Atlantic.

A joint expedition from Kew and Namibia's National Botanical Research Institute went to the Sperrgebiet in 2007 in search of, and to collect seed from, three of the area's most iconic species: *Welwitschia mirabilis, Aloe dichotoma* (quiver tree or kokerboom) and *Pachypodium namaquanum* (halfmens), and to see a coastal re-vegetation programme in action. A team of five, including Michiel van Slageren, the MSBP's International Coordinator for the area, set out from the coastal town of Swakopmund, heading eastwards for the famous *Welwitschia* desert. Here, this strange, slow-growing (it can live for more than 1,000 years) gymnosperm, which consists of two leaves and a massive taproot, is relatively common, but the Namib Desert (in Angola and Namibia) is the only place on earth where it grows. The team also came across the rare and threatened *Hoodia gordonii*; the Bushman candle (*Sarcocaulon marlothii*), whose stems contain a waxy substance that burns slowly when lit; and *Lithops ruschiorum*, which looks like a little brain lying on the ground.

From here the team drove to the Sossusvlei, the amazing sand-dune landscape of the Namib-Naukluft

RIGHT: Dune 45 in the Namibia's Sossusvlei National Park is one of the tallest dunes in the world. Climbing it yields collectable species such as *Sesamum abbreviatum*.

LEFT: The quiver tree,
Aloe dichotoma, in the field.
BELOW: Flowers and fruits
of *Aloe*.

National Park, part of the Namib, which is one of the oldest deserts on Earth. Here the towering, burnt orange dunes can be up to 300 m high. No Kew expedition had ever clambered up a massive sand dune to collect seed, but this time they succeeded in making a modest collection of *Sesamum abbreviatum*, a member of the sesame family that grows on the dunes.

After this, they headed across the Sperrgebiet, reaching the coast at Lüderitz, an old German mining town. Desert surrounded them, but they came across stands of succulent euphorbia shrubs and the occasional *Aloe dichotoma* (quiver tree). Today, the Namibian arm of the De Beers mining company, Namdeb Diamond Corporation (known as Namdeb), which is half owned by the Namibian government, has the concession for the whole Sperrgebiet and many of the mining villages of the early 20th century are abandoned. Deep in the restricted area, security is very tight, and the team frequently had to produce their papers for Namdeb officials so that they could continue their work.

Heading for Oranjemund, near the South African border, the route took them through a desert landscape made eerie by the occasional ghost town or small settlement. During the journey they kept a careful eye out for interesting species and potential collections, inspecting every possible site and making careful notes. They managed to collect some targeted species and noted some for future expeditions too. They finally reached the coast and completed the journey to Oranjemund on a coastal track, surfing perilously over the dunes that had blown across it.

They visited various sites along the coast where Namdeb's resident restoration ecologist and the MSBP are restoring areas disturbed by alluvial diamond mining. The coastal hummocks of saltbush, *Salsola nollothensis,* are crucial in the prevention of beach erosion, but they have been destroyed by mining activity. They are being re-sown, but are slow to establish. MSBP research shows that this is most likely due to seeds being buried too deep by unprecedented levels of shifting sand. Diamond–mining trenches channel the wind so that it gains in force and deposits sand swiftly. The MSBP's conservation team has developed a technique of forming piles of native seaweed around saltbush seeds to protect them from shifting sand. The hope is to restore the species on dunes along more than 600 km of Namibia's coast.

The expedition continued towards Sendelingsdrif, along the Orange River, collecting various targeted species as they travelled. Eventually, they left the Sperrgebiet near the Fish River Canyon, one of the main attractions in southern Namibia and often compared to the Grand Canyon. It cuts through stony semi-desert occasionally interrupted by green euphorbia or tall quiver trees. The MSBP team came across several target species in the area, including *Pachypodium namaquanum* (halfmens); unfortunately, although the extraordinary plants were in flower, there weren't sufficient seeds to make a collection.

On the way back, the expedition stopped for the night at Keetmanshoop. Near here they experienced one of the world's botanical wonders: a forest of quiver trees – *Aloe dichotoma* – in their hundreds, and protected too. Such is the richness and diversity of Namibia's drylands that the expedition came across all these botanical wonders in just two weeks. It's a botanical heritage that the MSBP is proud to help conserve.

Scouring Chile's hills for endangered botanical bounty

OPPOSITE: *Alstroemeria pelegrina* growing on the sea shores near Los Molles, Chile.

Those parts of the globe with a Mediterranean climate – wet winters and hot, dry summers – have a great diversity of flower species, and the Mediterranean parts of Chile are no exception. Chile is a long country, sandwiched between the Andes and the Pacific. To the north, the coastal regions are desert and include some of the driest places on Earth, while to the south grows lush temperate rainforest. The Mediterranean regions are found in the centre of the country and abut the montane and alpine flora of the Andes.

To a European eye, the plants are a mixture of the familiar and the strange. Species of buttercup, ragwort (*Senecio*), violet and spurge are common, while many plants that are widely grown in European gardens, such as alstroemerias, fuchsias, calceolarias and schizanthus, originate from South America. Less familiar are the many cacti and large terrestrial bromeliads (including *Puya*, which look rather like yuccas) that dominate much of the landscape.

Since Kew signed an agreement with the Chilean government's Agricultural Research Institute (INIA), Chile has become an important participant in the Millennium Seed Bank Partnership. I was fortunate enough to have the opportunity to join a field trip collecting seed for both the MSB and Chile's own seed bank. Pedro Léon-Lobos, INIA's project manager, and Michael Way, the MSBP's International Coordinator for the Americas, led the team of six.

The seed-collecting programme in Chile is well advanced, and most effort is targeted at particular species, especially rare endemics recorded at only a few sites. The team often locates a plant when it's in flower and notes its exact position using GPS (global positioning system), returning later when it has set seed. For our four-day trip we had a list of plants to look for in a region about 200 kilometres north of the capital, Santiago. We split our time between a range of hills and low mountains near the inland town of Caimanes, and a coastal strip of vegetation centred on the town of Los Vilos.

We got into the hills and mountains in rugged four-wheel-drive vehicles, using any track available, no matter how vertiginous. The steep hillsides were covered in low vegetation featuring numerous bulbous species, including several of our targets. In this terrain, collecting enough material for two seed banks is very time consuming and even with a team of six people, making four or five collections in a day is good going.

Sometimes we failed to make the collections we hoped for. One of our targets was a *Mutisia* species, a strange climbing member of the daisy family with stiff spiny leaves and showy red, orange or white flowers. Although we found many plants, insects had destroyed virtually all the seed.

Lunch consisted of the field-expedition staples of tinned fish and bread, supplemented with wonderfully tasty avocados. On the first day our fish attracted a juvenile Andean condor – these largest of all flying birds can smell carrion (and presumably tinned fish) from great distances. It circled us for ten minutes, coming quite low and was hugely impressive close to.

The vegetation in the valley bottoms is taller than on the hillsides, with many cacti, including branched candelabra-like species. Some of these were colonised by a curious species of mistletoe, its succulent berries dispersed by birds that perch on the cacti. The giant hummingbird, the largest in the world, is common here.

The coastal strip contains some of the most beautiful plant communities I've seen. These include wild species of *Alstroemeria*, which, although shorter than the familiar long-stemmed hybrids sold by florists, often have large flowers with the characteristic streaked nectar guides on their upper petals. We found six species and made several collections, including *A. pelegrina*, a seaside species that grows on cliffs and pebble beaches.

A little inland, we found two beautiful orchids of the genus *Chloraea*. Although distinctive in flower, in seed they are indistinguishable brown spikes. So we carefully marked 100 spikes of one species with inconspicuous stakes, allowing the team to identify and collect the seed on a return visit a month later.

The coastal plant communities are highly threatened by development. The beginning of the 21st century saw a boom in investment in hotels and seaside houses. Nowhere was this more obvious than at the pretty seaside village of Los Molles. Development had spread along the coast from the village, and, bizarrely, we saw endangered *Alstroemeria* species struggling for space among introduced pelargoniums and nasturtiums.

To the north of the village, development came to a halt at a wall, with the area beyond maintained as a reserve by an enlightened landowner. We had permission to enter the reserve, and the cliffs were smothered in colourful plants and cacti, including some very local species that we had targeted for seed collection. In places, the cacti form dense fields that are impossible to walk through.

The habitat's wildlife was fabulous. One islet had a colony of southern fur seals and an assortment of seabirds, including one that initially foxed me. It was standing at the top of the island, partly hidden by a boulder, looking a bit like a British guillemot. I was amazed, when it walked out from behind the rock, to see that it was a penguin – I'd forgotten that Humboldt penguins extended this far north, and hadn't realised they are mountaineers.

The Chilean biologists I met were wonderfully knowledgeable and committed to preventing Chile from decimating its biodiversity as has happened in Britain and the rest of Europe. Kew has a significant role to play in this movement by contributing to the development of seed banks and spreading conservation best practice.

Chapter 4

The science of saving seeds for posterity

In the bank: jars filled with seeds in the MSB vaults.

Inside the most biodiverse building on planet Earth

Nestled in the wooded Weald of Sussex, The Wellcome Trust Millennium Building (WTMB) is the Millennium Seed Bank Partnership's home. Tucked away from the hustle and bustle of London, the seed storage vault is arguably the most genetically diverse place on the planet. In the cold rooms beneath the main building there are 20,000 species in 48 m².

The vault is designed to last 500 years and it is undoubtedly one of the safest places in the south of England, with thick concrete walls. The cold rooms are underground, safe from dust and radioactivity, which is monitored via a sensor on the roof. If a nuclear power station, such as nearby Dungeness, ever exploded, the team at MSB could well be among the first in the southeast to know about it. Finally, to guard against floods, emergency pumps are on constant standby.

Yet the building that houses the storehouse for the world's seeds is far more than a secure repository, because it doesn't operate in isolation – quite the opposite. The Millennium Seed Bank acts as a hub for a burgeoning network of wild seed banks worldwide, many of which owe much to the support and advice of Kew's seed biologists and technologists. All over the planet, as habitat destruction continues, seeds from some of the world's most endangered plants are being conserved not just at Wakehurst but in their native lands, in seed banks that Kew has helped to inspire and equip. The circulation of knowledge and practical expertise through this network forms the lifeblood of the global project.

The Millennium Seed Bank building is a focus for vitally important research on the seeds themselves. Seed biologists at Wakehurst are not only discovering how best to store seeds safely for decades, even centuries, but also how to awaken these sleeping beauties again, so they can grow into vigorous plants for reintroduction to the wild. Often, staff at the MSB work with seeds that have never been handled before, and about which next to nothing is known. Seed banking buys time for environmentalists, and it is an ingenious and effective way of storing live seed and ultimately conserving large numbers of threatened plants. But even the most ambitious restoration scheme will be scuppered if no one knows how to store seeds so they remain viable, or how to germinate and grow on the world's wild seeds. So research into seed storage, dormancy and germination is at the heart of the project.

The curvaceous pod of the sea bean inspired the shape of the building, which was designed by London architects Stanton Williams and opened in 2000. It is named the Wellcome Trust Millennium Building in recognition of the two key sponsors: the Wellcome Trust and the Millennium Commission. It cost £17.8 million to construct, and in addition to the large underground vault for storing thousands of seed samples it includes advanced seed research and processing laboratories, an exhibition about seed conservation and the Millennium Seed Bank Partnership, 14 bedrooms for visiting scientists, a library and a seminar room holding up to 100 people.

Visitors walk into the glass-vaulted centre of the

OPPOSITE: The entrance to the MSB's underground seed vault.
ABOVE: Peter Randall-Page's 'Inner Compulsion' sculpture situated in the MSB grounds.

RIGHT: The MSB building in Sussex. Inspiration for the shape of the building came from the pod of the Sea Bean (*Entada gigas*).

OPPOSITE: Scarifying seeds to speed up germination.

building, which acts as a permanent exhibition space. Displays celebrate the diversity of the world's seeds, in all shapes and sizes, from the tiny dust-like orchid seeds to the 13-kilogram coco-de-mer or double coconut from the Seychelles, ten billion times the weight of an orchid seed.

The exhibition area is between the two research wings and through the glass walls visitors can see right into the laboratories where seeds are cleaned, processed and tested to make sure they are healthy enough to survive storage. Maps and photographs give a hint of the global connections at the core of the project, as does a tree-like sculpture made of business cards donated by scientists and conservationists from botanic gardens, universities, institutions and organisations all over the world. While visitors watch seed research and conservation in action in the laboratories, thousands of seeds lie stored in the underground vault beneath their feet.

Befitting its environmental role it strives to be a sustainable building. Recycling has been firmly entrenched at every level. The building is designed to maximise energy conservation, while providing the best possible conditions for seed storage. The cold rooms are chilled by a state-of-the-art system with many levels of control that help save energy. Waste heat from the unit is recovered for warming domestic hot water.

Inside, the laboratories' floors are lined with blue marmoleum, a natural hard-wearing covering fittingly made from flaxseeds. All the doors are yellow to complement the blue floors – a colour scheme chosen by the architect after the scientists insisted that the bright yellow 'hazardous waste' bins, standard in laboratories everywhere, had to remain on show.

Outside the building, the diversity of UK native flora is represented in eight parterres recreating habitats that include wetlands, downland, clifftops and a coastal beach.

One suite of laboratories is the domain of the curation team, who begin their work by sorting and cleaning the seeds as they arrive in a motley collection of bags, from the Sinai Desert, perhaps, or the mountains of Georgia. The drying room is nearby, and here seeds may spend several months slowly drying before they're ready to go to the underground vault for long-term storage.

Another suite of laboratories is home to the research staff. These scientists study dormancy and germination. They also troubleshoot germination problems that occur during regular tests, when samples of seeds are taken out of store and encouraged to germinate to check on their viability. The researchers also investigate new species as they arrive, to learn more about the nature and requirements of their seeds. They're investigating diagnostics too, discovering biophysical, chemical and molecular approaches to test or predict how seeds will behave in storage – and trying to understand how and why seeds age (see pages 134–7).

One wing of the building houses offices for staff, including the coordinators of the international programme. Downstairs, there's a meeting room, a library and guest rooms where visiting scholars and students can stay. Outside are the greenhouses, where a wide range of plants propagated from rare or endangered seeds are thriving – including the 200-year-old seeds once the property of Dutch merchant Jan Teerlink (see pages 120–3).

SEED BANK FACT

Safe storage

- The seed vault is almost five metres high and has a floor area of about 930 m².

- If filled wall-to-wall it could hold 100,000,000,000 rice grains or, at a tight fit, about 30 double-decker buses.

- The seed vault contains four 48 m² prefabricated cold rooms operating at -20°C. These are accessed from a drying room through an airlock. Fully fitted out, the vault could hold a further five cold rooms with the capacity for around 50 per cent of the world's seed-bearing species.

- Storing seeds underground reduces energy use in the summer and makes the vault more secure.

Laboratory and Building Manager

Managing the Wellcome Trust Millennium Building and its laboratories demands huge commitment, as the manager, Keith Manger, knows well. He's been at the Millennium Seed Bank since 1998, after a demanding career in both finance and the water industry as a commercial lab manager. 'I have designed labs and I have financial expertise, perhaps a unique combination,' he says.

He was just the man to oversee the building of the MSB in all its complexity, and to appreciate what has been achieved. Built on Jurassic sandstone, where Roman Britons once scraped a living, it's a site with a sense of enduring history. 'I've got a rock in my fish tank from under this building,' he says.

In 2000, Keith was among the 20 or so staff that moved into the new building. Now it has three times as many staff, plus an ever-changing complement of students and visitors. In such a complex operation, the manager's job is bound to be challenging, but Keith also has an eye on sustainability, and spends half his time giving technical advice to Kew's partners around the world.

'We're still seen as the mother ship in many countries,' he says, 'but we learn just as much from their experiences. The partnership is very much two-way.' For example, China's seed bank, which opened in 2008, is modelled on the Wakehurst design but has more advanced refrigeration technology. Seeing the benefits of the new system spurred the MSB to upgrade its machines. This move has made the MSB very energy efficient, as now the waste heat provides hot water for the visitors' accommodation.

While the MSB ticks over, much of the lab manager's time is taken up with helping MSBP partners all over the world design, equip and manage their own seed banks. 'We try to develop solutions that work for them, and are matched to their needs, but which leave the legacy of a network of national seed banks working to international standards.'

Keith Manger is working in China, South Africa, Botswana, Mali, Ethiopia, Burkina Faso, Tanzania, Kenya, Madagascar, Mauritius and more. 'Mali, for one, is a very successful project,' he says. Thanks to boundless local commitment and energy, the result of the partnership is a small single-storey building equipped with an MSB innovation: an inexpensive off-the-shelf incubator that acts as a perfect seed dryer. 'It's the ideal low-tech option that achieves international standards,' says Keith. Elsewhere, he laments, there are still places doing the wrong things, for instance storing seeds in plastic bags where they will simply die. 'It's not rocket science,' he says, ' but getting it right and sharing knowledge is what the MSBP project is all about.'

'We're still seen as the mother ship in many countries, but we learn just as much from their experiences. The partnership is very much two-way.'

The Millennium Seed Bank is situated within the grounds of the beautiful Wakehurst Place estate, in West Sussex

The journey of a seed though the Seed Bank

① Seeds arrive

Seeds arrive in the MSB from all over the world. They are usually in their fruits in cotton bags. They are put into a holding room.

③ Slowing decay

Most collections go to the dry room, which is the first crucial step to slow the rate at which seeds age and deteriorate. They begin the drying process in circulating air at 15°C and 15 per cent relative humidity; this is drier than most deserts. Seeds dry in approximately one month and are kept in the dry room for several months prior to cleaning.

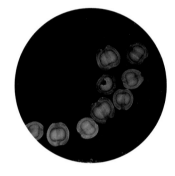

⑤ Is anyone at home?

A sample of the seed goes to the x-ray room where a seed morphologist checks it for quality – are there really seeds inside? Some seeds are too small to be x-rayed so they are cut in half and inspected under a microscope.

At this stage, some seeds go to quarantine to avoid spreading disease.

② Unpacking and checking

Seeds are unpacked in an isolation room and checked for live insects, damage and for storage viability. Any seeds containing live insects are separated and dried for a week in sealed cloth bags at -20°C to kill the insects. Then they can go through the normal process.

④ Time for cleaning

Cleaning involves removing twigs, leaves, flowers and other debris; extracting seeds from their fruits; and removing diseased and empty seeds. It lessens bulk and risk of disease. Dedicated teams of skilled workers clean the seed using equipment that includes sieves, and aspirators to winnow the debris away from the seed.

How many seeds?

The seeds are weighed to find out how many there are in the accession (a particular collection). An estimate is made based on weighing five samples of 50 seeds each to get an average weight and then weighing all the seeds.

Very small seeds are weighed in samples of 250. If an accession is very small, say 300 seeds or less, seeds are counted individually.

7

ID Parade

At this stage every type of seed that's stored is linked to a herbarium voucher (a dried specimen of the plant stored in Kew's Herbarium).

Records are kept of the seeds' progress at all stages.

8

Germination tests

Seeds are no use unless they will germinate and produce plants in the future, so a sample is tested to see if they are viable. Seeds are grown in Petri dishes or on agar plates and kept in an incubator at various temperatures. They are checked weekly.

9

Drying out

Before they can be stored, seeds are put back into the drying room at a controlled temperature and humidity, usually for less than a month, to reach equilibrium (no moisture in or out of the seed). The seed is tested for moisture using a hygrometer.

10

Packaging and storing

Packed into airtight containers before they leave the drying room, the seeds are then put into a cold room. The MSB uses glass storage jars with rubber seals that can be clamped shut. The cold store is kept at -20°C for long-term storage. Depending on the species, seeds may live decades, centuries or, in some cases, even millennia.

Preparing seeds for storage in the subterranean seed vault

There are four seed curation teams at the Millennium Seed Bank and their job is to look after the seeds throughout their stay in the vault. As you enter the Curation and Technology Section at the MSB, you can't miss the Seeds Arrival Board with its list of locations echoing an international airport: San Diego, Alice Springs, Lebanon, Malawi and Madagascar. Bags of seeds from these exotic destinations are allocated to one of four seed curation teams on arrival.

Drying is a crucial first step because it slows the rate at which seeds age and deteriorate. Seeds are usually dried in two stages to take them to a moisture content of about 5 to 10 per cent before banking. The first stage prepares them for cleaning, and often they are just left in their collecting bags in a drying room, a treatment that slowly draws water out of most seeds through their permeable seed coats. The drying room is kept at 15 per cent relative humidity and 15°C, which makes it drier than most deserts. When you stand inside, the evaporation off your skin makes it feel cooler than it is.

To enhance drying, air circulates all the time. Oxygen gets used up in the process, so there's a carbon dioxide monitor in the room that automatically brings in fresh air when oxygen levels fall dangerously low. Most seeds dry out after about a month. The seeds' moisture level is monitored with an electronic hygrometer – the biggest technical advance in seed science in the past decade. Before that, seed technicians had to judge degree of dehydration by drying a sample of seeds to destruction in an oven, and comparing their weights before and after.

When the seeds have dried they are cleaned, a meticulous process that involves separating the seed from all the plant stalks and flower heads. Sometimes, the curation staff use winnowers – machinery originally designed for the Dutch flower seed market – to agitate the plant material, causing it to separate from the debris. But the equipment doesn't suit many seeds, and much painstaking work is done by hand, in fume cupboards to keep dust away from technicians. After cleaning, seeds are dried again but before they are placed in the underground cold storage vaults the curation team nurture the seed (or a sample of it) through several more steps.

It is vital that every seed stored in the Bank is linked to a pressed and dried specimen of the plant from which it came. To make this happen, one member of every seed curation team liaises with the Herbarium at Kew. In this way, MSB staff can be certain that experts in a particular group of plants can always verify the identity of the seeds.

Before banking, a sample of the seed goes to the X-ray room, to check that there really are viable embryos inside. Sometimes seeds have insect damage but embryos may be missing from seeds for other reasons too. Some plants, particularly grasses and the daisy family, produce lots of empty seeds, perhaps to fool predators. If there are too many dead seeds it may not be worth putting the collection in the vaults, and knowing the proportion of viable seeds will help technicians predict how many seeds can be expected to germinate after storage.

Using a sophisticated digital X-ray machine, real-time images come up on screen, allowing the seeds to be viewed quickly at different magnifications and angles. It cost £70,000, but has already paid for itself in saved staff time. Some years ago, forensic scientists working

ABOVE: **Checking seed quality.**

OPPOSITE: **A visiting scientist conducting germination experiments.**

BELOW: Removing seeds
from a dried fruit head.
CENTRE: Seeds being put
into storage in the
underground freezers.

for the police used the machine to determine the age of insect larvae removed from a corpse to estimate the date of death; they were so impressed that they bought one for their own laboratory.

Some seeds, such as sedums, are too tiny to X-ray, so they're cut in half under a microscope to check their quality in a procedure known as an optional cut test. Next, samples of seeds are counted and weighed – a notoriously tedious task – then the whole collection is weighed so that the number of seeds it contains can be estimated.

Following this, samples from each collection are given an initial germination test, to check seeds are viable. This is a crucial step, as seeds are useless unless they can be germinated. Seeds are placed on Petri dishes or agar plates, kept at a variety of temperatures

in incubators and checked at weekly intervals for signs of germination. Some come up quickly like classroom mustard and cress, others can be quite slow and take months. If that approach doesn't work, then staff will try something more drastic, and consult a specialist in that family of plants. Over time, the curation teams have built up invaluable germination know-how; for example, members of the lily family with tiny embryos take longer to germinate than legumes do.

The job of the seed curators is still not yet complete. Before banking, a sample of seeds is tested to determine its likely storage life by being put through a rapid ageing process. This involves keeping seeds at high temperature and humidity, which kills them in a few weeks. At intervals some of the seeds from the test are germinated, and the results used to calculate

long-term survival rates in the ideal storage conditions of the seed bank. As a rule of thumb, storage life doubles for every one per cent drop in moisture content or 5°C drop in temperature. Germination tests are regularly repeated on batches of 50 to 100 seeds, usually at intervals of five or ten years, to check that seeds are surviving as predicted.

The MSB is a Defra-licensed quarantine centre and seeds must be put in quarantine if there's a risk of seed-borne disease: for instance, rhododendrons from the US might harbour sudden oak death fungal contamination (*Phytopthera*). To make doubly sure, plant material is never mixed with other waste; it goes into yellow bins and is then incinerated. Meanwhile, the rest of the batch is further dried and placed in containers before being taken for banking into the vault below the visitor's hall. A spiral staircase, visible from one end of the visitor's gallery, leads down to the door of the vaults.

Once properly dried, seeds are kept dry in glass storage jars. Every batch of jars has to be tested, to make sure they are air-tight, by adding a layer of self-indicating silica gel, then leaving them at 95 per cent relative humidity at -20°C. The jars are checked weekly. If the indicator doesn't change colour (from yellow to green) over the minimum period of a month, the storage jars pass the first test. Then they are kept at -20°C for another month as the final test.

At the moment, the MSB has three cold rooms in action; the fourth is empty ready for use. There is plenty of room for expansion by adding further cold rooms, potentially enough space to hold seed of 50 per cent of world's seed-bearing species. Inside it is cold, -20°C, which, with wind chill factored in, is equivalent to -27°C. Whenever a member of staff enters the cold room they must wear arctic gear – comprising blue coats, boots and gloves. When they enter, an alarm system is triggered and if they don't come out in ten minutes all hell breaks loose.

Paper work is a necessary evil for the team. Every shipment into the MSB must have all the relevant permits, such as CITES certificates, and as the seeds progress through the Curation and Technology Section everyone writes down what they do. In this way, knowledge is shared and passed on.

Finally, the curation teams also set to work whenever seeds are sent out to research labs or used for restoration projects. They don their arctic gear and venture into the vaults to retrieve seeds so they can warm up and prepare for life in the big wide world.

SEED BANK FACT

Save your own seeds

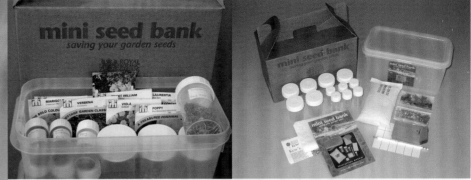

The MSBP's Mini Seed Bank is designed to enable the public to dry and store seeds. Relying on the same principles as the Millennium Seed Bank, the kit contains a polythene box that functions as a drying chamber and seed bank. It also contains a desiccant (drying agent) with coloured moisture indicator for drying the seeds.

The Mini Seed Bank is used by some of the MSBP's smaller partners, ensconced on islands with a small flora, or when seed collectors need to start drying seeds en route to the seed bank, via a long sea voyage or series of flights.

With conservation and seed banking now on the National Curriculum, a Mini Seed Bank has been sent to every primary school in the UK as part of the 2009 Great Plant Hunt project, to educate young people about plants and plant conservation. The inclusion of these subjects in the National Curriculum is a response to the global significance of habitat loss and species extinction. Mini Seed Banks have also been used in dozens of secondary schools, where A-level students have helped Kew scientists to investigate the longevity of Britain's wild seeds as part of the Save Our Seeds project.

LEFT: **The MSB Building.**
ABOVE: **The Mini Seed Bank Kit.**

A variety of neatly sorted and stored seeds.

John Adams

Technology Expert

The MSBP is at the forefront of developing new techniques for collecting, storing, studying and using seeds worldwide. With no precedent for much of its research, scientific and technological innovation are vital. The MSBP's technology expert and deputy lab manager, John Adams, not only looks after all aspects of the MSB building, from the heating and air conditioning to the security software and carbon dioxide monitors in the vaults (blow on them and you set off an alarm), he also develops new techniques that might be of use elsewhere. 'Problem-solving is something I like doing,' he says.

John has invented new and improved seed conservation technologies, including low-tech systems that are cheap and simple to use and will work almost anywhere. Whether it's finding a device to tell staff exactly how tight to screw up a bottle top, or quantifying the colour change in the indicator silica gel, John seeks out techniques used elsewhere in industry and science and applies them at the MSB.

He has tested how long it takes a jar of seeds removed from the cold room (-20°C) to reach room temperature. If you open a jar straight away, condensation forms inside, inadvertently wetting seeds that should be kept dry. The convention was to leave the jar for four hours before opening, but to find out the optimal time, he inserted tiny thermocouples attached to multi-channel data loggers that produce a temperature trace every five minutes. He discovered that it takes 13 hours for the biggest jars to reach room temperature. He advised staff to take jars out of the vault the day before they need to open them, to let them warm up overnight.

He also invented a drum-dryer for seeds. This uses an ingenious low-tech drying technique. Comprising a plastic drum with a secure lid, it has a central column on which you can hang seeds in bags above a desiccant in the bottom. You can tell the seeds are dry when the indicator silica gel inside the bags changes colour, from green to orange. Dozens of the drum-dryers are now in action around the world, from Eastern Europe to St Helena and the Ascension Islands.

If you don't have any indicator silica gel to hand, you can always use dry salt. Put the seeds in with dry salt and if the salt sticks to the glass, it's picked up water from your seeds, which are still too wet. Salt begins to clump at 75 per cent relative humidity. This easy test is fine for community seed banking, where the goal is to store the seed from one season to the next.

A second, more sophisticated, gadget comprises a plastic crate with a lid and a false bottom that has two holes drilled in it. In one hole you pop a second-hand computer ventilation fan that can run off a 12-volt solar panel or car battery. The fan circulates air over

SEED BANK FACT
The perfect jar

the desiccant lying in the bottom of the crate. The design gives a greater drying surface area than the drum, makes it easy to change to fresh desiccant, and is ideal for drying seeds inside a vehicle on collecting trips.

John has also discovered a brand of incubators that work brilliantly as seed dryers. They provide an air-tight chamber that will take up to a bucket-full of seeds (about 10 kg). Run at 18°C they produce an average relative humidity of 15 per cent. They even work in hot and cold greenhouses. No wonder these incubators have been shipped round the world, on the recommendation of the MSBP. They can serve as chambers for germination tests too, when they're not drying seeds.

John's latest plan is to automate seed cleaning by adopting the principle of gem-polishing drums. Could something like that work for some seeds, he wonders, if it were lined with rubber and you added rubber balls?

In the cold vault beneath the Wellcome Trust Millennium Building, around two billion seeds are stored in glass jars. Glass has the great advantage that you can see through it, but the opening must be sealed to stop moisture getting in. John has tested all the sizes and shapes of glass containers available, including second-hand marmalade jars from his local supermarket, which performed surprising well.

But it was granny's pickle jars – Kilner-type jars – with a rubber ring and metal clamp, which performed best. They are robust with a wide neck, stack well, and come in a range of sizes. A vital detail is that the sealing ring has to be natural rubber. The alternative, silicon rubber, may keep fish tanks and windows watertight but the MSB can't use it because it's porous (not moisture proof), a fact only discovered through extensive testing.

John still has to test each new batch of glass jars. First the jars are dried in the drying room then sachets of self-indicating silica gel are added. Orange indicates dryness, whereas green indicates the presence of moisture. A dry seed sample is sealed inside and the whole thing put into a very humid chamber at 90–100 per cent humidity for a month. If the seal works, the silica gel will stay orange.

For its partners, the MSBP recommends tri-laminate foil bags, which comprise a military-grade layer of aluminium foil with plastic either side. Once heat-welded, you can stand on one full of water. They seal and pack very well and come in various sizes, so fit neatly into boxes and make the best use of storage space. The MSBP itself has opted for glass so the seeds inside can be seen – making their work, literally, transparent.

LEFT: **The ultimate storage** solution: small glass bottles filled with seeds inside rubber-sealed clamp-top jars.

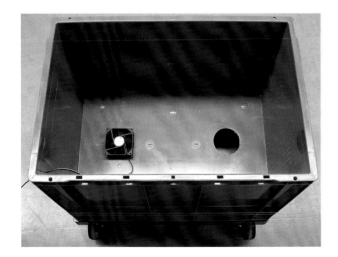

Cutting-edge seed research

Humans could not have spread across the globe in such numbers had we not perfected ways of collecting and storing the seeds of wild plants. Our very ability to grow our own food – now transformed into a huge global industry – owes its origin to the discovery, millennia ago, that the seeds of most crop species can be dried and preserved in a viable form from year to year.

Now seed conservators are working to expand humanity's knowledge far beyond domesticated crops, to discover how to store seeds from those wild species about which, so far, virtually nothing is known. Researchers are finding ways to judge which newcomer seeds can be safely stowed away in a seed bank, and to ensure they are given optimal treatment while they're being conserved for the future.

At the Millennium Seed Bank, research scientists and seed technologists are at the cutting edge of seed conservation. Not only does their work involve storing seeds, it also includes advancing understanding and perfecting germination of stored seeds; there's little point in storing seeds if they can't be germinated and grow into viable plants. Their work includes predicting how long seeds can live (seed longevity); the quest to discover how to get dormant seeds to spring into life (germination prediction); how to deal with seeds that cannot cope with being dried (recalcitrant seeds), as drying is an essential step to successful storage; understanding how seeds respond to the seasons (seasonal adaptation) and research into how and why seeds age. All these areas of research are essential to increase humanity's ability to bank seeds for future use.

Seeds of the traveller's palm *Ravenala madagascariensis* have a beautiful bright blue outer coating.

Unravelling the mystery of how long seeds live

An important part of the seed conservationists' knowledge is to know how long the seeds from a particular wild species can live in seed bank conditions. Such information is well established for our agricultural and horticultural species. For example, we know that onion and lettuce seeds have relatively short lifespans. With many agricultural species, it's better to buy or collect fresh seed every year or two rather than hanging on to packets of seeds for years and years. By contrast, radish and runner beans produce much longer-lived seeds.

But when it comes to wild species, for the most part, we simply don't know which wild plants produce seeds that are inherently short- or long-lived. This gap in our knowledge is a concern for the world's seed banks, all of which routinely remove stored seed at fixed intervals – every five or ten years – to check their viability through germination tests. These fixed intervals may be too long for short-lived species, and needlessly frequent for long-lived ones.

It is only now that Kew's seed biologists are finding ways of distinguishing the long-lived 'Methuselah seeds' from those that are destined to be short-lived. Intensive research began five years ago by looking at seeds stored in the MSB from 195 representative wild species drawn from 71 of the world's plant families. Their native habitats ranged from cold deserts to tropical forests, arrayed across the globe from Finland to Chile and from New Zealand to Canada.

MSBP seed scientists compared the longevity of different species by the technique of artificial ageing. During these tests, a sample of seeds is kept in hot, moist conditions: 45°C and 60 per cent humidity. This treatment speeds the natural ageing of seeds,

because it accelerates the reactions that cause viability loss. To discover what proportion of seed is still viable, a few seeds are removed from the incubator at regular intervals and given the right conditions to germinate. The research revealed that the seed itself can often provide the first clue to its longevity. In endospermic seeds, the bulk of the seed interior is filled with the seed's food store, the endosperm, and when they leave the parent plant the

RIGHT: An accelerated ageing experiment.
OPPOSITE: Exploring why a seed has failed to germinate.

embryos need to develop further before the seeds are able to germinate. Seeds with small embryos relative to the size of the seed – so called endospermic seeds – proved to be significantly shorter-lived than seeds without endosperm. The research also revealed that seeds from cooler, wetter climes tend to be shorter-lived, whereas those from hot, dry regions are more likely to be long-lived.

Consistent with this finding, complementaty research has found that seeds of alpine species are consistently shorter-lived than related species from lowland areas. These findings support recent theories that the first flowering plants evolved in moist habitats such as montane forests and riverbanks. These plant pioneers produced seeds that didn't need to survive for long periods in a dry state. Longer-lived seeds probably evolved later, when flowering plants began to invade regions with harsher, drier climates. Currently, Kew's scientists are now analysing the germination data of stored MSB seed collections spanning the last 40 years to shed even more light on how to predict seed longevity.

Thanks to the new research, seed biologists can begin to identify short-lived species more rapidly than before. Replenishing banked seed of short-lived species from genera such as *Primula*, *Gentiana* and *Rhododendron* may mean re-collecting seeds from the wild more frequently or regenerating seed by growing up stock plants at more frequent intervals than before. At the other extreme, exceptionally long-lived seeds – such as some of the Myrtaceae species tested at the MSB – can be re-tested less frequently, say every 25 years, thus saving valuable seeds and staff time.

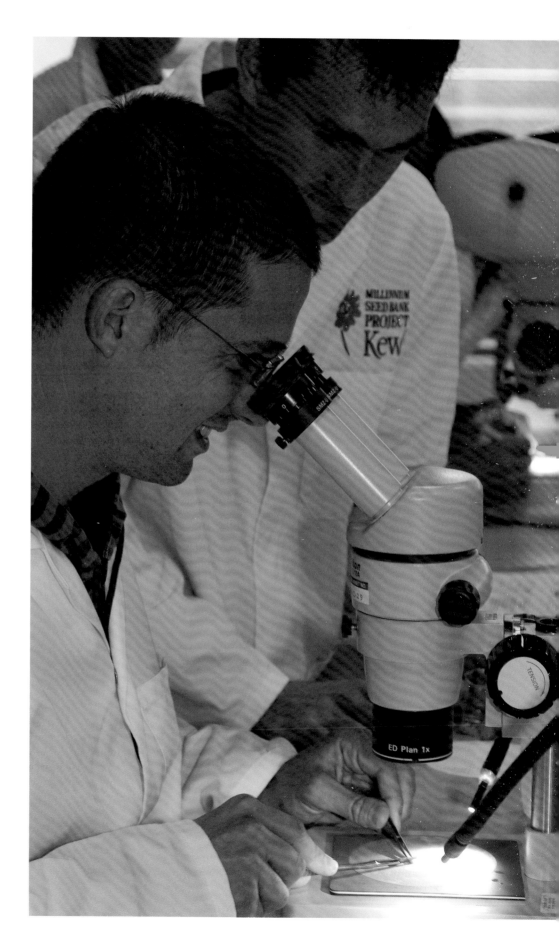

RIGHT: *Ammocharis* sp.
looking splendid in full bloom.

The seed that would not dry

When a new seed from Namibia arrived at the MSB it went into the drying room, as seeds normally do, but it didn't dry out. A species of *Ammocharis*, related to our daffodils, it looks like a corm, or bulb, but is actually a seed with a thin, corky seed coat that proved incredibly resistant to letting water out. In fact, it germinated soon after it arrived, forming a bulb, root and shoot without any water and was still alive in the drying room two years later. A native of arid deserts, and possibly even a new species to science, it evidently found the dry room very much to its liking.

BOTANIC BOOTY...

DAVID RUMSEY MAP COLLECTION

SOUTH AFRICA: The *Henriette* lies at anchor off Cape Town, her hold crammed with silks, porcelain and tea. It is March 1803 and the ship is bound for the Netherlands from Java and China, but she has broken her voyage of many months to take on supplies before rounding the Cape of Good Hope and sailing north for home.

1803: In England, George III is on the throne; it is two years before the Battle of Trafalgar; Britain is at war with France, and the Netherlands is an annex of Napoleon's France.

The story of seeds that germinated in the MSB in 2006 began during King George III's reign.

Off Cape Town, the Dutch merchant Jan Teerlink leaves the *Henriette* and goes ashore to stretch his legs, explore and socialise before the last leg of the voyage home. At some point during the two-week break he visits the renowned Company Gardens, planted some 250 years earlier by the Dutch East India Company and filled with plants of the famed fynbos flora of the Cape.

LEFT: George III.

Teerlink acquires precious seeds from Company Gardens and carries them back to the *Henriette*. In all he has 40 folded paper packets of seeds which he tucks into his red leather wallet for safekeeping. Then, as now, the Netherlands was a nation of garden lovers, mad for plants. These rare seeds from the Cape were a considerable prize, possibly destined for a botanical garden or the private collection of a wealthy Dutchman.

The *Henriette*'s papers are still among the High Court of Admiralty's 'prize papers' – documents seized from captured ships – when they are moved to the UK's National Archives at Kew, in the 20th century. And here they might remain for decades more but for the Dutch historian Roelof van Gelder, who embarks on a research project for the Royal Dutch Library in 2005 and comes across a red leather wallet with the words 'Jan Teerlink of Vlissingen' (Flushing) embossed in gold across the front. Inside are 40 packets of seeds.

Whatever their destination, the seeds do not reach it because the British capture the *Henriette* sometime during the last days of her voyage between Cape Town and the Netherlands. She is so very nearly home. They seize her valuable cargo, and send all her papers, including Teerlink's red leather wallet, to the Tower of London for safekeeping. As for Teerlink? He is released and goes home empty handed. He had lost his precious cargo and important personal papers, but not his life.

ABOVE: The wallet of Jan Teerlink, filled with small bags of seeds.

RIGHT: The 200-year old seeds Teerlink gathered in South Africa.

But is this really botanical treasure? Not unless the 200-year-old seeds can be germinated, and this seemed highly unlikely. Although ancient seeds have, on extremely rare occasions, been encouraged into life, it is almost impossible for seeds to remain viable for so long in less than perfect conditions.

...SPRINGS TO LIFE

UK: Van Gelder wants to try. He eventually sends samples of the seeds to the Millennium Seed Bank, where seed scientists reveal that the 40 packets contain 32 species of seed, including legumes, *Erica*, daisies and members of the *Protea* family. They are all plants of the fynbos.

ABOVE: **King protea,** *Protea cynaroides*, **flower.**

ABOVE: **Coaxing the seeds to germinate required careful investigation of their favoured habitat and conditions.**

To get the seeds to sprout, the scientists have to unlock the germination codes and this requires knowledge on what it takes to get seeds from the Cape's dry Mediterranean climate, with its frequent wildfires, to germinate.

For many fynbos seeds, germination is triggered by the effects of wildfire so MSB scientists simulate that by wetting the seeds with water through which smoke has bubbled or by chipping the seeds' tough outer coats with a scalpel to simulate cracking in the intense heat of a fire.

MARCH 2006:

Of a sample of 25 legume seeds, 16 germinate and prove to be *Liparia villosa*. They are followed in June the same year by an *Acacia* seed, and in July by one of eight *Leucospermum* seeds.

'You might not think three species out of 32 is very impressive, but I'd have been happy if just one seed had grown,' says Matt Daws, seed ecologist.

None of the *Liparia villosa* seedlings survive. But at the time of writing the single acacia is growing well in the seed bank glasshouse at Wakehurst Place. It is about a metre high now, but has yet to flower so its identity remains a mystery. The *Leucospermum* is now a healthy-looking, bushy specimen that's grown to about 0.8 metres high.

Leucospermum – a healthy-looking, bushy specimen that has grown to about 0.8 metres high.

Seedlings of *Liparia villosa* (left) and *Acacia nilotica* (right). It is hard to believe that this healthy plant grew from 200-year-old seeds.

What does the future hold? Do these extremely rare examples of their species shed any scientific light? Scientists hope so. They have taken cuttings of the *Leucospermum* to produce further plants for research. They plan to compare them with present-day examples and hope to gain important information about the differences between species that grew 200 years ago and those that grow today. The treasure hunt continues...

The quest for the germination predictor

Getting wild plant species to germinate is often easier said than done. Finding out what triggers germination can be tricky as, depending on the species, triggers can vary from extreme fluctuations in temperature to smoke and fire.

Research scientists at the MSB constantly monitor the seed collections through routine testing. They focus on those species that have failed to germinate despite repeated attempts using different conditions. Many plant species use dormancy, a protective measure to ensure that seeds will not germinate unless conditions are perfect for good seedling growth.

Dormant seeds are time travellers: they wait for the right time and place to germinate. For example, seeds shed from a plant in late autumn might soon experience a severe drop in temperature as winter approaches, making this a bad time for a tiny seedling to grow. To coax a living but dormant seed to germinate, MSBP researchers such as Lindsay Robb have to know what conditions a particular seed needs to break its dormancy. If they don't know, they have to make an educated guess and experiment by offering germination conditions that relate to the habitat and climate from which the seeds originate. 'Nature has designed these seeds so that they can function perfectly in their environment, so it makes sense that we look at that environment in order to understand how to germinate them in the laboratory' says Lindsay.

The MSB research team has a goal: to create a global 'germination predictor', a comprehensive tool that can tell anyone exactly what they need to do to get any seed in the world to germinate. They are working towards that goal using the UK flora as a testing ground. The quest is to discover the simplest germination regime that works. It is detective work with ecological intelligence at the fore. For example, seeds from wetland habitats, such as rushes and sedges, often need alternating day versus night temperatures for germination. This ensures seeds only germinate at the margins of water bodies, not submersed at depths from which they would not be able to grow. Knowing when and where seeds germinate in nature provides vital clues to seed detectives.

Tetrazolium stain testing is a powerful tool that can be used to understand seed quality, germination and longevity. Add tetrazolium chloride to prepared seed and if their metabolisms are active, they should turn red. That's the theory, but in practice, what you often see in wild seeds are various shades of pink to red and patchy staining. It takes a seasoned eye to judge a seed's viability, which can then be used to validate germination test results and verify the quality of a seed collection.

Some seeds are very awkward to germinate; members of the rose family, for instance, are notoriously difficult to coax into life. Dunking rose hips in concentrated sulphuric acid in order to simulate the action of a bird's digestive juices can work, but it is dangerous work and not reliable. Research is underway to discover whether nitrate might be a key, as it is with some other hedgerow species: perhaps it's not transit through the avian gut so much as ending up in the bird's dropping that spurs rose hips to germinate.

OPPOSITE: **Fire is used to break the dormancy of seeds of** *Grewia bicolor*.

Head of Seed Conservation and Technology

The MSBP's Head of Seed Conservation and Technology, Robin Probert, says he became 'hooked on seeds' when he was an undergraduate in London and worked at Wakehurst Place as part of his degree course. He joined the staff two years later in 1976. For the next 24 years he concentrated on research, investigating the germination, dormancy and storage of seeds; research aimed at underpinning the conservation of wild plants through seed banking.

As the longest-serving member of the team, he remembers when Kew's seed bank was still a fledgling. In those days, long before the Wellcome Trust Millennium Building was established, the then-modest collection of seeds was lodged in the former chapel at the Mansion House at Wakehurst Place in Sussex. Thankfully, he says, in the early 1990s Kew's trustees decided not only that Kew should continue to bank seeds, but that the project needed to be drastically expanded, so that it could do more than merely scratch the surface of a significant global problem.

Once the MSBP was established, Probert and his team acted as the interface between research and the running of the seed bank itself. He became an expert at what he calls 'troubleshooting', solving germination problems, and finding ways to crack seed dormancy. 'It's detective work,' he explains. 'We have to think, what is this plant, what climate has it evolved in, when might it germinate in the wild?' Sometimes, when next to nothing is known about a particular species, the team uses the location data logged when the seeds were collected. This enables them to access a climate profile from the WorldClim database, which gives patterns of rainfall and temperatures. 'We can then set up a simulation to replicate wet and dry seasons and try to

unlock the seed that way.' He cites an example from America in which MSBP scientists working on the germination of a North American wild species of the carrot family were unable to germinate seeds using their normal protocol. They knew that the seeds had been collected in Utah, and a check of the climate data revealed that the seeds hadn't been chilled sufficiently to simulate Utah's long, cold winters.

Probert and his colleagues have shed new light on seed dormancy as an adaptation to seasonality, as the story of the wood anemone reveals (see page 129). 'This kind of research shows how finely tuned plants are to seasonal climates,' he says, 'and hints at how profoundly climate change will affect many wild plants.'

He has also made a major contribution to our understanding of how long seeds will survive, both in nature and in the seed bank. The scientific study of the comparative longevity of seeds is quite new, and still poorly understood. Probert and his colleagues have discovered that seeds from cooler, moister regions with tiny embryos are inherently short-lived, compared to seeds from warmer, drier climes and with bigger embryos inside. 'Climate of origin and seed structure are two key predictors of seed longevity,' he concludes. Such knowledge will enable seed curators all over the world to be better informed about how best to manage their collections.

Understanding how seeds rely on seasonal shifts

A surprising discovery at the Millennium Seed Bank led to new insights into the wood anemone *Anemone nemorosa* and highlighted its vulnerability in the face of climate change. Alarm bells went off when staff doing routine checks on banked wood anemone seeds found they would not germinate. After just one year of storage, they were dead. Collected from local Sussex woodland, they formed part of the MSBP's landmark project to bank seeds from every native UK flowering species.

It soon became clear that the seeds hadn't survived the standard drying regime, the vital first step in preparing seeds for long-term storage at the MSB. This was a surprise because, for most seeds, gradual drying is the best way to shut down their metabolism and slow ageing. In that suspended state, many seeds can remain viable in the deep freeze for decades, even centuries.

So why wouldn't the seeds of wood anemones tolerate that initial drying? The MSB team suspected the answer lay in the natural life history of this denizen of ancient woodlands, and they devised ingenious experiments to tap into the wood anemone's reproductive secrets. Researchers put freshly harvested seeds in nylon mesh bags and popped them back into the leaf litter on the floor of the wood. Every month, a sachet was retrieved and examined, while temperature changes were monitored via a remote data sensor, also buried in the leaf litter.

The results revealed that when the wood anemone sheds its seeds in late May or early June, they are still immature and the embryos inside are tiny. As the summer progresses, the embryos develop while the seeds lie safely nestled in moist leaf litter and this process makes all the difference to their bankability. Unlike most seeds shed by the mother plant, which have to cope with drying out, wood anemone seeds do not have to withstand desiccation, so they have never evolved the capacity to cope with drying out. When autumn comes and the temperature falls, they begin to germinate by sending out pioneer roots. But the shoots themselves remain dormant until they have experienced a period of winter cold. Then, in spring, warmer temperatures are the trigger for the shoots to begin to emerge.

It turns out that wood anemones living high up in the Apennine Mountains of northern Italy behave in a similar fashion, but each stage in a seed's life is exquisitely adapted to suit the local climate. Mountain wood anemones do everything at much cooler temperatures, and require longer winter chilling before their shoots will sprout, so they wait till the snow melts.

MSB scientists are concerned that, as wood anemones are so finely tuned to their local seasonal climate, climate change is bound to disturb their reproductive events. Seedlings may struggle to become established as springs warm up, and competitors may swamp what is presently one of our earliest spring flowers. Although abundant today, the wood anemone could become vulnerable in the future.

Meanwhile, will it ever be possible to successfully bank their seeds? The team at the MSB says that it is all about timing, and points to a narrow window of opportunity. Their research shows that in the first couple of weeks after dispersal, the developing embryos briefly become tolerant of drying and could be stored in the bank, but as development continues, they lose this ability. All the same, it still looks as though the seeds are inherently extremely short lived, designed by evolution for a brief but vital stay in their damp leaf litter.

OPPOSITE: *Anemone nemorosa* in full flower in Bethlehem Wood, Wakehurst Place.

To dry or not to dry? Understanding recalcitrant seeds

In most cases, drying a seed is the best way to shut down its metabolism temporarily and so extend its ultimate lifespan. By delaying germination indefinitely, drying slows the ageing of seeds, ensuring their longevity sometimes for hundreds or even thousands of years, a very important advantage in seed banking.

Seeds that cope well with drying are said to be desiccation-tolerant, or orthodox. Such seeds are often able to survive in the soil, where they form a natural seed bank, patiently waiting year after year until conditions are right to germinate. It is thanks to this natural seed bank that red poppies suddenly spring up in a newly ploughed field: they have germinated from seeds lying dormant in the soil.

However, not all seeds can withstand drying – such seeds are said to be desiccation-sensitive, or recalcitrant – and they are difficult to store for anything longer than the short-term. For example, seeds of the mango, *Mangifera indica*, die after relatively soon drying. More than 500 plant species with recalcitrant seeds have been identified to date but estimates suggest that at least eight per cent of the world's flowering plants, may share this trait.

Seed bankers need to know, or be able to predict, which seeds are likely to be desiccation-sensitive, and therefore difficult to store. Research by the MSBP has discovered that recalcitrant seeds tend to be larger and rounder than average, weighing more than half a gram each. This shape and size reduces the rate at which they lose water. Instead of biding their time in the soil, these seeds are intent on rapid germination. When they are shed from the mother plant, recalcitrant seeds are primed for germination: metabolically active and with a high moisture content. Because seedlings are less likely

to be eaten than seeds, rapid germination is a good way to avoid seed predators. On average, recalcitrant seeds also have thinner seed coats. The seed coat might account for just 15 per cent of seed mass, compared to 46 per cent in an orthodox seed. These seeds don't waste resources building a thick seed coat, as there is no point in investing in a stout defensive shield if you're not planning on hanging around long enough to be gnawed by a hungry rodent.

It turns out that just two seed traits can predict the probability of a woody species having desiccation-sensitive seeds, and so forewarn seed-bankers. These are seed mass and seed coat ratio (thickness). Not surprisingly, species that adopt the rapid seed germination strategy tend to be native to tropical forests that are reliably wet: Palms, are an example. A few trees in the tropical drylands also produce recalcitrant seeds, which are released just as the annual rainfall reaches its maximum. A good example is the oily seed of the shea butter tree, *Vitellaria paradoxa*, an endangered African species highly valued for its fruit and timber.

Even temperate species may produce recalcitrant seeds. The English oak, *Quercus robur*, does just that. Its seeds are recalcitrant and also have delayed germination or dormancy built in. Acorns need to wait until spring to germinate. They survive the winter thanks to seed predators, who bury the recalcitrant seeds in the moist woodland floor. Luckily for the oaks, squirrels and jays never manage to retrieve and eat all the acorns they have buried.

Provenance and a seed's developmental history can also influence whether seeds survive drying – as work at the MSBP on sycamores and horse chestnuts has revealed.

ABOVE: **Deliciously juicy mango (*Mangifera indica*) dangling from a branch.**

AMAZING PLANT FACT

Great Greek conkers

The horse chestnut, *Aesculus hippocastanum*, was introduced to Britain from the Balkans in the 1500s. Today, horse chestnut trees in Greece produce seeds that are, on average, five times larger than those produced by trees further north in Scotland. Many Scottish conkers probably never mature completely, and as a result they may struggle to germinate and be less able to survive periods of drought.

LEFT: **English oak** (*Quercus robur*).
BELOW: **Leaves and flowers of the Egyptian thorn tree** (*Acacia nilotica*).

AMAZING PLANT FACT

Thorn in the desert

Seeds of the Egyptian thorn tree *Acacia nilotica* can withstand drying until they are virtually water-free, down to less than five per cent moisture content. A mature tree can produce 2,000 to 3,000 pods in a good fruiting season and each contains eight to 16 seeds. The tree, which is widespread in Africa, can withstand extremely dry environments and floods. Introduced in Australia, it has become an invasive species and is regarded as a pest plant of national concern.

Hugh Pritchard

Head of Research

The Head of Research at the Millennium Seed Bank is a cryobiologist by training, a specialist in storing seeds at very low temperatures. But over his long career at Kew Hugh Pritchard has studied every aspect of seed biology, and is, he says, a jack-of-all-trades. With teaching links at the University of Sussex and the University of Bedfordshire, he supervises PhD students each year, edits scientific journals and has an honorary professorship at China's leading botanical institute, an affiliation that reflects his key role in the founding of China's national seed bank for wild plants.

An internationalist at heart, Pritchard stresses the MSBP's role in forging links throughout the world. 'I like to see these engagements with the outside world as a series of concentric circles,' he says. The key partner countries form the core, boosted by seed donation programmes with many other countries, such as Canada. Beyond, the MSB is scientifically engaged with many more nations, and the first links with people and countries are usually through research and universities.

Pritchard emphasises that the MSBP, as part of Kew, is a biodiversity institute. 'In some ways we're similar to universities, in other ways we resemble independent research institutes or industry, all at same time,' he says. 'We're in a unique position to translate fundamental understanding into applied outcomes because we sit in the middle: universities and industry are encouraged to produce products for sale, but we have a different outcome.'

The challenge confronting Pritchard and his fellow seed scientists is the realisation that there are risks attached to handling seeds from species that the world's scientists know little or nothing about.

'We need to ask, how can we predict the performance of the seeds? It's not enough just to wait.'

In search of answers, the team follows two broad approaches to research: first, there's the more practical, immediate questions when seeds come in, are banked, and then there's a problem: that is when the technical or applied side of the research swings into action. Secondly, there's the search for general, basic understanding, when the researchers devise broad hypotheses and then test them by drawing on material held in the MSB.

Fundamental research focuses on searching for markers that will enable curators to screen seeds, assessing their viability, for instance, using very few seeds. Finding ways to overcome seed dormancy – and knowing which approach to use with which seeds – is vital too, as is judging what seeds will survive drying, which is the vital quality for storage. 'We are looking for patterns across diverse biological families, trying to understand nature in ways that will help us to predict how seeds will behave, now and in the future,' says Hugh.

Climate change makes understanding dormancy and germination increasingly urgent. 'Seeds are absolutely important to the future of plants,' he says. 'Seeds are the main means of regeneration in the natural world - they have memories, they are living - and the way they respond with us is relevant to how they'll behave in their natural setting.'

Searching for the elixir of plant life

Why do seeds die? The short answer, of course, is that nothing lives for ever, but some seeds live for centuries, while others struggle to survive a year. If we understand how seeds age, we might be able to assess how long particular seeds can be stored and remain viable. Insights garnered from such research could also shed light on human ageing, because the fundamental molecular processes at the heart of life and death are widely shared across the living world.

At the MSBP, biochemists are looking for diagnostic markers of seed viability, so that they can recognise senility in seeds. Of course, seeds that are dead or dying will mostly fail to germinate, but some species naturally remain dormant for many months before they germinate, so germination tests aren't always a reliable guide to the health of a consignment of seeds.

The MSBP team has discovered what looks to be a universal marker that can measure the degree of stress within seed cells, and so predict whether the seed is doomed to die soon. The marker is an ubiquitous antioxidant, called glutathione. The researchers found that seed viability decreased by 50 per cent when the antioxidant capacity of glutathione reached a key threshold, a signal that the cell is damaged beyond repair. At that tipping point, this signal acts as a 'death trigger' forcing cells to initiate a self-destruct programme, and if many cells self-destruct the seed will die. The team tested representatives of 13 plant and fungal orders, and found that in each case the marker was an accurate indicator of plant-cell viability.

When seeds are profoundly dry their metabolism is virtually at a standstill. But not entirely, as some genes are still active in very dry seeds. The MSBP team is pinpointing them, and trying to understand how seeds continue to age even in the deep freeze. They suspect that in such conditions, normal gene regulation may be bypassed, with biochemical reactions happening spontaneously via molecular 'nano switches'. Such processes could explain why seeds continue to age even when they are thoroughly dry and cold.

Finding these genes and understanding how they work is a huge challenge. Research on human cells is easy as they are relatively simple structures. Seeds, however, are packed with a vast array of carbohydrates, lipids and proteins designed to nourish the germinating seedling. They also have chemical defences. For example, seed coats are often laced with phenols that play havoc with conventional biochemistry. It can take months of work in the laboratory just to sort out the right procedure for triggering germination.

Left over seeds in your garden shed are ageing as you read this, but seeds buried in the soil, awaiting the right moment to germinate, can extend their lifespan every time in rains. MSBP researchers have discovered that seeds can survive year after year in the soil because they are able to repair DNA whenever the soil becomes wet. Similarly, in the laboratory, each time seeds experienced simulated rainfall they produced the ubiquitous antioxidant glutathione, which mops up reactive and damage-causing free radicals that have built up as they have aged in dry soil.

This remarkable phenomenon means that seed bankers may be able to improve the longevity of aged seeds by periodically taking the seeds out of dry storage and re-hydrating them using high humidity for a brief period. This technique, known in the horticultural trade as priming, has long been employed to stimulate rapid germination in commercial seed stocks.

OPPOSITE: Charlotte Seal using the MSB's fat analyser. Knowing the precise oil content of seeds is important for estimating viability and longevity.

Green chemistry

Sunflower, rapeseed and soya bean seeds are just some of many seeds that are rich in oils. Throughout much of Africa, the oil extracted from the shea butter tree, *Vitellaria paradoxa*, is used to protect and moisturise the skin. Research by the MSBP is analysing these seed oils to find new oilseed species suitable for sustainable agroforestry or agriculture. They have perfected a way of extracting oils without using highly toxic solvents. Instead, they use non-toxic, inexpensive and non-flammable carbon dioxide.

Now we know that it can also increase the longevity of some seeds. Its rejuvenating effect works best on seeds that have already spent some time in store. The team found that priming foxglove seeds that were not yet fully ripe encouraged them to mature and subsequently boosted their longevity in storage.

In tropical forests up to half of all seed-bearing plants produce recalcitrant seeds that cannot be dried for conventional storage. Research is under way to find ways to isolate and dry just the embryo itself, which could then be stored at very low temperatures in liquid nitrogen. Isolated embryos turn out to be more tolerant to dehydration than intact seeds, and the trick is to dry the embryos just enough to remove the risk of ice crystals forming inside cells and rupturing them.

It is a delicate operation, and success depends on an intimate understanding of the embryos' response to the whole procedure. Together with colleagues in Russia and South Africa, MSBP biochemists have discovered that when the embryos are cut away from sweet chestnut seeds, they release a burst of superoxide, which acts as an antiseptic, sterilising their surface to ward off invading microbes. Our bodies do something similar when we cut a finger. In both cases, the surge of superoxide also acts as a signal to the immune system, inducing a defensive response and mobilising wound-healing genes. In the future, cryopreservation techniques might be improved by modulating the superoxide burst, to stimulate or suppress it. For example, applying antioxidants to the excised embryos could prove useful through dampening down any adverse effects of an over-vigorous superoxide burst.

The smell of distress

Just as human breath can sometimes be analysed to detect disorders, researchers at the MSB are investigating the volatile molecules given off by seeds for clues to their state of health. When seeds are under stress, molecules in their cells begin to break down and release volatile hydrocarbons. Using laboratory techniques such as gas chromatography-mass spectrometry, these compounds can be detected in samples of air above stored seeds. The quest for seeds' volatile fingerprints is exciting but challenging, as each species releases a distinctive cocktail of compounds when under stress.

ABOVE: **The recalcitrant seed of *Vitellaria paradoxa*, source of shea butter, cracked open.**
OPPOSITE: **Louise Colville operating the gas chromatography-mass spectrometer.**

Chapter 5

Sharing knowledge and putting seeds to work

Cleaned and labelled seed
collected in Burkina Faso
and Mali.

Exchanging seed secrets with global partners

Equipping people around the world with the skills to collect, store and germinate seeds from rare, threatened and useful plants is a vital part of the MSBP's role. Since 2000, MSBP staff have trained more than 1,500 people in seed conservation skills. Trainees include 16 foreign students from eight countries who have completed PhDs, and 19 students from 14 countries who are presently undertaking their doctorates. Pretty much everyone working at the seed bank has an aspect of training to their job.

'It's a question of combining our partners' expertise with our expertise,' explains Kate Gold, training manager at the MSBP. 'They obviously have a better idea than we do about the plants and flora of their country but might be less knowledgeable about how to collect seeds of wild plants. There are always gaps in knowledge and understanding that need to be filled. We start by finding out what people already know, then work out what's missing and how can we address that.'

The types of training vary widely. The MSBP runs a three-week Seed Conservation Techniques course at Wakehurst Place every two years, which is designed to cover everything students need to know about working on a seed conservation project. Many partner organisations have sent members of staff to attend this course. In other cases, people come and spend up to a month working alongside seed bank staff. They gain practical hands-on training that they can put into practice back home.

MSBP staff also visit partners abroad to conduct training in host countries. This can be particularly beneficial as trainees work with equipment and plants they are familiar with, and more people can attend sessions. For example, 22 people attended workshops

on Seed Conservation Techniques at the South African National Biodiversity Institute (SANBI). The courses included two days of theory on seed collecting and handling techniques and one day collecting in the field. A similar course was held at the Kunming Institute of Botany at the launch of the MSBP's partnership with China.

Training courses generally comprise a mixture of lectures with PowerPoint presentations, and practicals conducted in the field and laboratories. MSBP staff also provide supporting materials to help trainees. These include Collection Guides explaining where botanists might find targeted species, what they look like and when they might be ripe and therefore able to yield seeds. Downloadable technical information sheets are also available, covering all aspects of seed conservation, from how to select containers for long-term seed storage to the best ways of designing seed-drying rooms.

ABOVE & RIGHT: **Training at** the MSB and in the field.

The MSBP tries, where possible, to assess the effectiveness of its courses. It provides all trainees with a questionnaire asking what aspects of their course they thought were effective and what elements could be improved. When staff visit projects abroad, they assess how partners are putting into practice what they have learned in training. There is an emphasis on trainees passing on their new knowledge and skills to others. 'The team at KIB translated our materials into Chinese and have now trained more than 200 new seed collectors,' says Kate.

The success that partners have had in meeting their own seed-gathering targets reflects the progress the MSBP has made, in its first decade, in passing on its seed-conservation practices. In four years, the Tasmanian programme has secured 930 seed collections, including 138 threatened taxa; the MSBP partner in Burkina Faso has also made over 900 collections, representing 60 per cent of the nation's flora; and, in South Africa, SANBI collected seeds from more than ten per cent of its flora, some 2,500 taxa. The latter included seeds from 55 species classified as critically endangered.

Livhuwani Auldrean Nkuna of South Africa testifies to the helpfulness of the MSBP's training programme. 'The course covered almost everything I needed. I am now able to plan very successful seed-collecting trips. I am able to make good-quality seed collections due to the fact that I know the right time to collect seed from a wide variety of targeted species. The course boosted my confidence in liaising with landowners to gain access to collect seeds and in forming partnerships and collaborations with conservation organisations. All the people I met were friendly and helpful and I had so much fun,' he says.

ABOVE: Inspecting seed quality during a training course.

Vanessa Sutcliffe

Technology Specialist

OPPOSITE: **Vanessa Sutcliffe** collecting seed in Abu Dhabi.

How best to bank seeds is a vital part of the MSBP's training programme. From Burkina Faso to Kyrgyzstan, the MSBP is at work spreading knowledge of the techniques of seed banking throughout the world. The Technology Specialist gives advice to other partner countries, to help them develop facilities and successful techniques. 'But', warns post holder Vanessa Sutcliffe, 'it's vital to be very open minded, not trying to stick to a rigorous schedule. You need to be prepared to go with the flow, and give help and support where you are needed.'

The job is very varied. Vanessa helped to train staff in China's new seed bank in Kunming. 'Everyone was really keen to learn as much as they could, and to understand each process,' she says. 'In their vault, they've used all the technology we've used, it's like walking into the one at Wakehurst ... they've really taken our advice.' In Georgia, giving lab-based training to staff, she remembers the ready humour and warm, insistent hospitality of the partners. In Kyrgyzstan, she trained seed collectors in the field and taught them how to look after their collections after harvest. Everywhere she goes she's impressed by commitment to the common goal to conserve the world's irreplaceable wild plants.

Back at the MSB, the job includes producing the technical information sheets that have proved so useful for burgeoning seed banks. Small-scale seed drying methods and low-cost monitors of seed moisture status are just two of the recent additions to the series. She's also designing posters on subjects such as seed cleaning, which seed banks can stick up on their walls for easy reference.

As Technology Specialist, Vanessa also played a role in a series of workshops for African seed banks, supported by the United Nations' Food and Agriculture Organisation (FAO). The workshops focused on species that seed banks were having difficulty storing or germinating – so called 'difficult' seeds. The 'Difficult' Seeds Project' is funded by the UK's Department for Environment, Food and Rural Affairs and helps farmers and seed banks in Africa to conserve plants used for food and agriculture. As part of the project Vanessa led training workshops for seed technicians: two conducted in French in Burkina Faso and Morocco, and two in English in Kenya and Botswana.

At each workshop, a dozen or so technicians were brought together to follow a programme of training in seed conservation techniques, and were able to share their experiences and develop friendships that forged links between national institutions. Two-day sessions for farmers were organised too, to help smallholders improve their storage of crop seeds or overcome problems with germination. The project website contains species profiles which allow any gene bank to find advice on how to deal with seeds from some 160 species that seed conservators have found difficult to handle, store or germinate. 'It helps the seed banks that have participated in the workshops, but also casts our nets much wider,' says Vanessa.

For more on the 'Difficult' Seeds Project see pages 180–181.

Kate Gold

Training Manager

Kate Gold explaining the workings of a psychometric chart during a training course at the MSB.

Training is an integral part of the MSBP's international programme and grows expertise around the world. Training Manager and seed biologist Kate Gold is ideally placed to help MSBP partner countries fill the gap between what they want to do and what they need to know to succeed.

Kate organises and helps to teach a vast range of courses at the MSB and abroad. Practical hands-on work is mixed with theoretical learning, and involves everything from collecting seeds to storing them. She's already worked in most partner countries, including the USA, Chile, Kenya, Burkina Faso, South Africa, Botswana and Australia.

The goal is to promote best-practice seed conservation across the globe. Kate stresses how vital it is to share understanding of the principles behind seed conservation within the constraints and limitations of any particular set-up. You don't need fancy kit. 'Seed conservation isn't rocket science, the most important part is drying seeds adequately, then storing them in air-tight containers, and cooling them,' she explains.

Kate is also helping communities overcome problems in using seeds effectively. 'If there's a problem with germination, say, we can help.' It might be that nicking seed coats or dousing seeds with hot water or acid will do the trick. It's often a matter of trial and error. 'It's important that the technique is practical, cheap and not too time consuming. I can bring that perspective, and search for low-tech solutions,' she says.

By training trainers and making technical know-how available on both information sheets and the internet, the goal is to help partners build their capacity to the extent that they no longer need any back-up from the MSB. This has already been achieved in many places.

China's Germplasm Bank of Wild Species (GBWS)

The role that seed scientists can play as ambassadors of goodwill between nations is illustrated by the development of the Germplasm Bank of Wild Species (GBWS) in China, which evolved in part through a fruitful ten-year collaboration between Chinese seed biologists and the MSBP. It has been built in south-western China, a region of exceptional plant biodiversity that is increasingly under threat from agricultural and industrial development.

Chinese scientists made their first formal visit to Wakehurst in November 1999. This was followed by a series of summer visits by specialists from Yunnan province. When Kew's MSBP formally joined with the Chinese Academy of Sciences in 2004, one of Kew's key aims was to help the Chinese partner accelerate its wild plant conservation work. The new centre opened in October 2008, with six senior members of the MSBP staff present at the opening. Seeds from 204 species of UK flora are now lodged in the Chinese facility.

MSBP staff helped the Chinese design their facility, which is bigger than the Wakehurst building and different in layout. But the cold and dry rooms look identical, and were built by the same company. MSBP staff also helped to train the Chinese staff, both at the MSB and in capacity-building workshops held in China. The head of the seed bank operation at the new genebank, Dr Xiang-yun Yang, completed her PhD at the MSB in 1999.

Joint scientific initiatives are now under way, with two PhD studentships and the development of a seed germination testing programme at the GBWS. The Chinese seed bankers plan to collect and store seeds from 10,000 threatened and endemic plant species from China by 2015.

Sharing seed knowledge boosts local livelihoods

BELOW: Kate Gold shares experiences with women from a community in Kenya.

Training and capacity building are central to the MSBP's Seeds for Life project in Kenya. This nine-year project is working with five partners: the National Museums of Kenya; the Kenya Agricultural Research Institute (KARI, represented by the National Genebank of Kenya); the Kenya Forestry Research Institute (represented by KEFRI, Kenya Forestry Seed Centre); the Kenya Forest Service; and the Kenya Wildlife Service. As well as strengthening these partners' abilities to find, collect and conserve seeds of targeted species, the aim is to work closely with local communities and help farmers exploit useful species in a sustainable way.

Around 7,500 plant species grow amid Kenya's drylands, moist tropical lowland forests and equatorial mountains. Of these, around 1,100 are endemic to the country. Urban developments by the country's growing population, slash-and-burn agriculture and over-grazing have reduced wilderness areas in the past

30 years, fragmenting habitats. Forest cover has dropped from ten per cent to two per cent during the same period. Droughts also frequently take their toll on the nation's flora. Saving seeds from plants before they become extinct has therefore become a priority.

Farmers in Kenya are keen to plant trees to help reverse deforestation, provide them with timber, fruits and nuts and to secure water supplies. However, it is often difficult for them to obtain seeds. Kenyan partners' staff trained local communities in the basic principles of seed collecting, processing, germination, nursery techniques and management. As a direct result of this work, several new species were collected, described and banked; foresters, farmers and botanists began working together and demonstration plantings helped rehabilitate degraded habitats around the districts of Mbeere and West Pokot.

Members of the partner institutions also came to

Soundtrack to seed collecting

the MSBP, where they researched techniques for germinating and storing Kenyan orchid seeds and African tree seeds. They used the knowledge they gained to develop and deliver an undergraduate diploma course in seed conservation techniques at Maseno University. Three Kenyans, meanwhile, undertook the Kew Plant Conservation Techniques Course. They contributed to the Seeds for Life project by creating a standard operating procedure for examining seeds using X-ray at KEFRI and producing a species prioritisation list.

As a result of the Seeds for Life Project, some 2,500 seed collections, representing nearly 2,000 Kenyan plant species, are now in safe storage at seed banks in Kenya and the MSB. Seeds from around 2,500 more species could be collected relatively easily. Plants saved for posterity include *Aloe ballyi*, a tree listed as endangered in the 2010 IUCN Red List of Threatened Species. Populations of the species, which is only known from five locations in Kenya and Tanzania, are decreasing because the dry bushlands it favours are exploited for charcoal production. One known population has recently been lost as a result of a road-widening scheme.

Another locally vulnerable plant now banked is the desert rose *Adenium obesum*. This succulent shrub or small tree, often swollen at the base, has become a victim of its own toxicity. So poisonous that birds cannot nest in its branches, the desert rose is frequently destroyed where it grows. However, local Tharaka people use it for its medicinal properties. The bark of the plant is chewed to aid abortions, while powdered stems are rubbed on livestock skins to control fleas and lice. Toxic plants often contain compounds that are effective as pesticides or medicines, so future research on the desert rose and many other species could yield further valuable uses.

A casual glance through the pages of *Kew Magazine* inspired London-based music producer Dan Massie to use music to bring environmental messages to young people. He approached Paul Smith, Head of Seed Conservation, with the idea of producing a CD of songs with a conservation message based on the MSBP's work. Paul put him in touch with the MSBP's International Coordinator for East Africa, Tim Pearce, who suggested recording an album in Kenya, drawing on the talents of local musicians.

The pair headed to Africa in 2008, where they recorded everything from ancient chants to Swahili hip-hop in hotel rooms, school halls, Kenya National Theatre's rehearsal rooms and memorably in the 'bush studio' of the open countryside. The result is an album *Seeds for Life,* that includes rap artist Ukoo Flani singing 'we're living in nature, you cut down trees, bury your nation' and Kenyan superstar Teddy Kalanda Harrison urging 'seeds bring life, seeds give life, seeds for life.'

In the Taita Hills, Dan recorded tribal music including traditional drumming accompanied by local women singing and dancing. The women's lyrics urge the need to balance human livelihoods and the environment. They are well aware of the need to do so, as the large trees required to make quality drums have long been cut down; today the drummers stretch animal skins over oil drums instead. It is a good illustration of the need to restore forests and an affirmation of why projects such as Seeds for Life can bring real benefits to African people.

ABOVE: **Kenyan musicians contributing to the *Seeds for Life* album.**

Where are they now –
former students around the world

'I attended the course while working for the Eden Project in Cornwall. It made me realise we had serious seed-storage problems, as our seeds were kept in plastic bags and cardboard boxes in cupboards. After undertaking germination tests, we removed dead or poor viability seeds then dried, packaged and stored the good seeds using techniques demonstrated at the MSBP. I am now passing on the seed-banking skills I learned to horticultural students on placement at the Eden Project.'

MAUREEN NEWTON, UK
Seed Conservation Techniques course

'Supported jointly by the MSBP and the Kunming Institute of Botany, my PhD project is aimed at understanding why seeds die. I am using pea (*Pisum sativum*), tea (*Camellia sinensis*) and sweet chestnut (*Castanea sativa*) seeds to unravel mechanisms that cause cells to die during storage and desiccation. These plants have, respectively, orthodox, intermediate and recalcitrant seeds. I hope my work will help us better understand cell death and lead to enhanced protocols for seed storage in China and South-East Asia.'

HONGYING CHEN, CHINA
PhD student, Seed biochemistry and molecular biology

'I work with the Kenya Wildlife Service, one of the partners of the Kenya Seeds for Life project. I am responsible for generating and making available, in an easy-to-understand format, scientific information that will enhance the management of Kenya's natural resources, both fauna and flora. The eight-week PCS course covered plant conservation options, from managing protected areas and botanic gardens to seed banking and cryopreservation. It has really widened my skills. Back home I am applying the skills I acquired to help conserve and promote the sustainable use of genetic plant resources.'

JAMES MATHENGE, KENYA
International Diploma in Plant Conservation Strategies (PCS)

I work for the MSBP's USA partner the Lady Bird Johnson Wildflower Center (LBJWC). I attended the MSBP's three-week Seed Conservation Techniques Course in 2004 and have since been involved with training volunteer seed collectors back in the USA. Prior to the course I was new to the seed-collecting game and so learned a lot. I had passed up many collections thinking the populations were not large enough. Since the course I have gone back to those populations and made several collections. After completing the course we were able to develop a 'mini' seed conservation workshop for our volunteers at LBJWC. They love this programme as they feel they are making a difference and enjoy the opportunity to travel to different parts of Texas to collect seed.'

MICHAEL EASON, USA
Seed Conservation Techniques course

'I work for the Western Australian Department of Environment and Conservation as a technical officer in its Threatened Flora Seed Centre (TFSC). I began a part-time MSBP-supported PhD in 2005 to address concerns that commonly used *ex situ* storage conditions might not suit orthodox Australian seeds. I have now completed a review of the TFSC's collection re-test data, highlighting the successful application of seed storage to the Western Australia flora. I have made a number of research visits to the MSBP, including one to model seed viability of *Xanthorrhoea preissii*, a critically endangered Australian tree with grass-like leaves at its base.'

ANDREW CRAWFORD, AUSTRALIA
PhD, Seed longevity and viability

'A University of Illinois-Chicago student, I work with the Chicago Botanic Garden and the MSBP. I am examining the consequences of non-random seed harvesting techniques across the fruiting season in a North American prairie species. One of my findings is that early fruiting plants have seeds with rare 'gene flow' events. This is where pollen donors are from outside the population as opposed to more common gene flow between plants in close proximity. This demonstrates that if seed collectors do not collect from early fruiting plants, they could miss rare genetic material. The MSBP provided funding and intellectual support; as part of my research I spent six weeks at the MSB studying seed dormancy.'

JENNIFER ISON, USA
PhD, molecular genetics

SID shares data with others

No sooner have MSBP staff collected, described, photographed and germinated seeds from new species that arrive daily at Wakehurst Place, than information garnered during these procedures is compiled and disseminated to the wider world. This is done using a vehicle known as SID, or the Seed Information Database. Essentially a digital seed bank containing information on some of the seeds in the physical seed store, it is placed online so that anyone, from academics and government officials to seed companies, can query the data.

'The database grew from an earlier compendium of published literature on seeds that were difficult to store,' explains John Dickie, the MSBP's Head of Botanical Information. 'Kew staff needed information that would help them build up pictures about individual species, generate hypotheses for testing and develop lines of enquiry for future research and decision-support. Today, we accumulate information from published papers as well as taking unpublished data from external contributors. We combine it and make

it available to anyone who wants to use it.'

The International Plant Genetic Resources Institute (IPGRI) and the University of Reading's Department of Agriculture collaborated with Kew to provide seed information for the initial compendium; MSBP staff have subsequently expanded the database over the past decade. As of March 2010, SID contained records for 33,346 taxa (the word taxa describes a taxonomic group of any rank, such as species, family or class). Information is included, where available, on: storage behaviour; seed weights, dispersal methods, germination, oil content and protein content; plant life-form; seed morphology; salt tolerance and seed viability.

The database is simple to use. A quick search for the Olive tree *Olea europaea*, for example, reveals it has orthodox seeds (those that can survive drying and freezing) that yield an average oil content of 20.97 per cent. The olive tree is well-known for its economic value, providing edible fruits, sought-after wood, and high-quality oil for cooking. As well as providing academics with useful data for scientific research,

the characteristics of seeds contained in the database can help guide researchers to little-known species that could be similarly useful.

SID could not exist without input from other departments at Kew. For example, the taxonomists in Kew's Herbarium play a vital role in ensuring plants are correctly named. Kew is currently working to assign all names at generic level and below to the families they are assigned to in the latest classification system, known as APG III. 'Synonomy, where one species has more than one name at various times in its life, can make querying databases quite difficult,' says John. 'We are constantly in touch with colleagues to ensure we use correct botanical information in SID.'

SID is part of wider efforts at Kew to digitise its historical data and combine it with contemporary records. Efforts to link various databases mean researchers can now gather together information on all aspects of a plant if need be. The initiative to bring botanical data together in this way is known as the electronic Plant Information Centre (ePIC) project.

Searching ePIC for *Olea europaea* reveals that, alongside five records within SID, there are 39 specimens in Kew's living collection, 8 references in the Economic Botany bibliography, and 51 references in the Library Catalogue, among others.

Scientists at the MSBP and Kew use the growing array of information available to them to help partners, who often have few botanical resources, target seed collections in their home countries. This involves taking information from SID, such as whether seeds can be easily stored or not, and combining it with geographic information collected from specimens at Kew and elsewhere about where populations of particular species are known to exist, and biological information such as when plants are likely to set seeds. The information is presented to partners in the form of 'Collection Guides' so they can efficiently prioritise and plan itineraries for seed-collecting expeditions. Guides recently completed include *Trees of Botswana*, *Threatened Taxa of Albany* (South Africa) and the *Malawi Red Data List*.

Growing useful plants in villages around the globe

Throughout the world, small communities rely directly on wild plants for their livelihoods. Yet species that are considered very useful are often those most at risk of being lost through over harvesting or human development. The Useful Plants Project aims to help local communities store and propagate species of wild plants, which they select as being of particular use. This may be as food or medicine, or as building materials or fuel. Sometimes a single species of wild plant fills all those functions: cacti with edible fruits often have many valuable uses, as do palms, acacias and scores of medicinal plants chosen as priority species by the local communities. The Project is managed by Tiziana Ulian, from Kew's Seed Conservation Department, through partnerships in Botswana, Kenya, Mali, Mexico and South Africa.

Each community targets its own array of plants to suit local needs and priorities. The project is designed to build capacity in whatever ways are appropriate. Often, the wild species have never been propagated before, so MSBP input can help overcome problems with germination and ensure that seeds are stored in the best way possible. It's all about knowledge transfer to the local community. Help might be working out how best to germinate seeds from a useful plant, or finding appropriate storage solutions for small farmers, so they can store seeds on the farm. A bucket in the sun may not be the best way to keep seeds but there might be ways to modify traditional methods to increase the seeds' longevity.

The first step is to identify the target species selected by the villagers, and then to collect and store seeds, ready for sustainable use. Seeds are stored in the country of origin and duplicates of every collection are stored in the MSB in the UK. The project supports propagation of these valued plants in nurseries or community gardens where they can be readily distributed to local people. In some collaborations, scientists in partner institutions also carry out research into the eco-physiological requirements of the plants, or into the phytochemistry of medicinal properties. In South Africa, new greenhouses built in the Lowveld Botanical Garden are being used to grow useful species, and provide training to local people.

The number of species targeted varies from country to country. In Mali, where people's everyday lives are particularly closely linked to plants, 200 species are under investigation, and the project is already making an impact, as nurseries are established in rural communities and plants grown on in a plot donated to the community by a local farmer. Seven rural communities are taking part in the project, and seed-collecting teams from the Institut d'Economie Rurale in Sikasso have already collected and banked seeds from many key species. Meanwhile at the National Public Health Research Institute in Bamako, Dr Rokia Sanogo is investigating traditional medicinal plants, some of which are processed and marketed for sale.

Useful Plants Project
safeguarding plants for local communities

CLOCKWISE FROM TOP RIGHT:

Traditional healer with school child, Botswana; Workshop teaching plant propagation techniques, South Africa; Interviewing a traditional healer, Kenya; Sharing expertise with the community in Kenya.

The Useful Plants Project (UPP) runs in countries where many inhabitants depend directly on natural resources for food, medicine, fuel, building materials and income. So far, under the careful leadership of Tiziana Ulian, the project has propagated 585 useful species as well as developing and sharing knowledge on the best possible propagation procedures, helping local communities to improve their livelihoods and use the natural resources in a more sustainable way. Here are some examples of amazing plants safeguarded by this project.

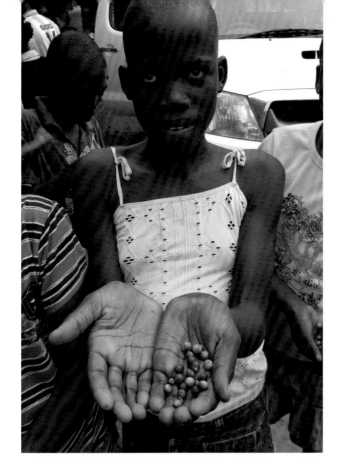

CLOCKWISE FROM TOP LEFT:

Schoolchild showing fruits of *Grewia flava*, Botswana; Fruits of *Ximennia caffra*, Botswana; *Xanthophyllum gilletii*, Kenya; A planting event in Tsetseng community, Botswana.

KENYA

Zanthoxylum gilletii (African satinwood)

This fine timber tree is often planted in compounds due to a traditional belief that the trees help keep away evil spirits. Its main traditional use is as a medicinal plant. Although the seeds are used to spice teas, which are drunk to relieve coughs, colds, flu and chest ailments, the bark is the usual source of medicinal compounds. African satinwood is used to treat rheumatism, malaria and cardiac conditions as well as gastro-intestinal infections. Scientific research into *Z. gilletii* has revealed that the bark contains chemical compounds which have a potent anti-microbial activity, supporting its medicinal uses.

Kigelia africana (sausage tree, cucumber tree)

The common names for this tree reflect the distinctive shape of its large fruits, which can weigh up to 12 kg. The main traditional uses of this tree are medicinal. There are many different applications. It is used to treat blood, digestive and nervous orders as well as infections, inflammation, poisoning, respiratory problems and skin complaints. There is strong scientific support for the medicinal activity of chemicals present in extracts from the roots, bark, leaves and fruits. *Kigelia* can now be found in European skin care products, such as those produced by African Earth and Zambesia Botanica, and may be of particular benefit for people with sensitive skin, eczema or psoriasis.

BOTSWANA

Grewia flava (velvet raisin, brandybrush)

The fruits of *Grewia flava* are of great importance to Botswanans living subsistence livelihoods, providing both calories and much-needed income. The revenue derived from the sale of the fruits is relatively high, possibly a reflection of the fruits' use in beer production, including the illicit *khadi*. The fruits have more flesh than other *Grewia* species and are eaten in large quantities, providing essential calories and vitamins. The flesh is also mashed, soaked and eaten as porridge.

Ximenia caffra (large sourplum)

The ripe fruits have a refreshing sour taste and are a popular food, considered best when slightly overripe. The fruit can can be preserved as a jam or jelly, or it can be dried in the sun. Dried fruits are sometimes added to porridge to increase flavour and content. The kernel of the stone is edible and has a 65 per cent oil content, leading to its use as a skin-softening cosmetic, among other things. The leaves and roots are used in a variety of traditional medicinal preparations including as a remedy for dysentery and diarrhoea, for fever, for morning sickness and for infertility.

RIGHT: Display of fruits of useful plants, Mali.

BELOW LEFT: Display of useful plants at Koutiala in Mali.

BELOW RIGHT: Rokia Sanogo, Mali.

MALI

Psorospermum febrifugum

In traditional medicine, *Psorospermum febrifugum* is used to treat skin conditions, but scientific research is indicating possible new uses for this species. Despite testing negative against malaria and avaian malaria, a number of new compounds with medicinal properties have been isolated from it and are currently being investigated. *P. febrifugum* is also a priority species for study in relation to its interaction with antiretroviral therapies offered to HIV-positive patients in Uganda, where the plant has become endangered. Thanks to UPP *P. febrifugum* seeds have been collected for banking in Mali and at the MSB, and planted in three newly-established medicinal gardens.

Lippia chevalieri

Traditional uses of this medicinal plant include using the flowers to prevent headaches and employing leaf decoctions as painkillers and to treat pulmonary and stomach troubles. Scientific tests have indicated that the plant's essential oils have a general antibiotic action. The most important use of *L. chevalieri* in Mali is as a constituent of the 'Malarial 5' treatment, produced by project partners the Department de Medicine Traditionelle. Malaria rates in Mali are high: with the disease affecting an eighth of the population it is the country's leading cause of mortality and morbidity.

MEXICO

Stenocereus stellatus (Xoconostle)

The beautiful columnar cactus *Stenocereus stellatus* is prized in central Mexico for its edible fruits, which are harvested from plants growing in the wild as well as those on cultivated land. Mature fruits contain a sweet, juicy pulp full of sugary liquid, which can be fermented to make a potent drink called *colonche*. The seeds are sometimes eaten or else roasted and ground with chillies in sauces. If the spines are removed, the young branches can be used as goat feed. The plants are used as living fences and as soil stabilisers on cultivated slopes.

Cyrtocarpus procera

This small tree with very pale grey bark is native to thorn-scrub forest in Mexico. Fragments of seeds from the species have been found inside the Coxcatlan Cave at the very earliest levels, suggesting humans were using the fruits as food in 4500 BC. The trees are still in cultivation in the Tehuacan Valley, and also grow wild as small groups or scattered individuals. In season, the fruits are sold in local markets. Earlier research into medicinal flora of the San Rafael region identified *C. procera* as an important treatment for kidney ailments. Through the UPP the species antimicrobial activity is being investigated further.

SOUTH AFRICA

Eucomis autumnalis (pineapple flower, pineapple lily)

Eucomis species are widely used in South Africa as traditional medicine. This particular species is made into decoctions to treat urinary diseases, and is the second most popular species sold in the Durban medicinal trade. However, this trade is not sustainable and collectors are having to travel greater distances to more remote locations to collect bulbs to sell. The trade removes 428,000 complete plants each year. Over recent years travel times have risen by 100 per cent from 1–4 hours to 3–8 hours. Unsurprisingly, *Eucomis autumnalis* is now judged to be vulnerable to extinction in the wild and has been placed on South Africa's Red Data List.

Trema orientalis (pigeon wood)

The common name pigeon wood indicates the tree's ability to attract large flocks of pigeons: a desirable trait as it allows the birds to be easily hunted as food. The tree can survive long periods of drought. Subsistence farmers use *T. orientalis* in a number of ways. The bark makes an excellent string, and can also be used for dyes and waterproofing for fishing lines. The leaves are eaten as a vegetable and used as animal fodder. The wood is mainly used for furniture and shelving. The fruits and leaves are used medicinally to make teas to treat respiratory disorders such as bronchitis, pneumonia and pleurisy.

Restoring habitats in South Africa

ABOVE: *Erica verticillata*, a species from the sand fynbos habitat.

OPPOSITE, CLOCKWISE FROM TOP LEFT: Carly Cowell and team planting near Cape Town; planting in the Bracken Reserve; restoring Tokai-Soetvlei burn near Cape Town.

The Cape Flats sand fynbos habitat once covered the southern tip of South Africa, but as Cape Town expanded to become the country's second most populous city, this lowland habitat shrank to nine per cent of its former extent. In 2005, a collaborative project between the MSBP and South African National Biodiversity Institute (SANBI) set about restoring remaining patches of the endangered habitat and re-establishing it in areas from which it had disappeared. The benefits of this work are clear to see at the longest-standing restoration site, Kenilworth Racecourse, where restored fynbos is attracting monkey beetles, sugarbirds and Cape sun birds to the area for the first time in a century.

Coordinator for the Cape for SANBI, Carly Ruth Cowell, initiated the restoration work at Kenilworth Racecourse. A field assessment revealed that although the site was one of most pristine remaining areas of fynbos, it was missing some key species. 'We found some of them in the living plant collections at Kew Gardens, some at Vienna Botanical Gardens, some at Kirstenbosch National Botanical Gardens (KNBG) and some in the Millennium Seed Bank,' explains Carly. 'We grew them up and then planted them out at Kenilworth. However, because we didn't know which plants liked which conditions, we just had to plant them in rows and see which ones survived. Over the course of two years, we established exactly where each of the plants likes to grow.'

This experience is now being put to good use at another site, Tokai Plantation. A working pine plantation covering some 1,000 hectares, the site is being reallocated for restoration in blocks, as two-decade-old pines mature and are felled for timber. This habitat was nowhere near as intact as the Kenilworth Racecourse

site, so the team began by burning the land due to be restored. Because the proteas, restias, ericas and bulbs that make up the lowland fynbos vegetation are responsive to fire, the aim of this was to trigger any natural seeds lying dormant in the soil to begin growing again. 'We have quite a few coming up and we're now identifying what's missing with the aim of tracking those down and planting them at the site,' says Carly.

To help with future restorations, Carly and her team have begun building up seed stocks, with collections divided between seed banks in the government's Agricultural Department in Pretoria, KNBG and the MSB. They are using the 'Purcell List', a list of plants encountered on the Cape Flats by amateur botanist Dr Purcell in the 1800s, as a reference for trying to gather seeds of all the plants that represent the Cape Flats' sand fynbos habitat. 'We've now got most of them in the bank but we suspect some of them may be extinct,' says Carly. However, it's still possible that some of those rare plants may be out there somewhere, waiting to be found and their seeds banked for posterity. You can read the story of how lost lowland fynbos species *Erica verticillata* was tracked down on page 45.

The Kenilworth site is now sufficiently established to be providing a source stock of plants for transferring to other restoration sites. With the habitat returned to its former glory as far as possible, the owners of the site, Gold Circle Racing, have employed a full-time conservation officer to maintain the vegetation. 'The racecourse owners are very enthusiastic about what we're doing and have always given us access to the habitat,' says Carly. 'It's a lovely model to take to other companies to encourage them to look after the tiny bit of natural vegetation that they may have on their land.'

Restoring habitats in the Middle East

With dryland seeds some of the easiest to bank for posterity, the MSBP has built up an extensive stock from Middle Eastern plants, along with knowledge about their germination requirements. Meanwhile Kew's Herbarium is home to taxonomists with expertise on Middle Eastern vegetation, along with botanical records made by naturalists, including Sir Wilfred Thesiger, at times when habitats were intact. These resources are now being put to use in projects to restore degraded habitats in the United Arab Emirates and Kuwait.

In one project, Kew has been advising architects who are landscaping an archaeological site and installing a visitor centre, on how best to restore vegetation. The plants that grew in the location in Bronze Age times are similar to Eastern Arabia's natural vegetation today. "We've given them advice on what species and plant communities to use," explains Shahina Ghazanfar, a taxonomist at Kew with expertise in Middle Eastern botany. "The site has sandy desert soil that is partially saline. There are two or three species of trees and large shrubs that suit these conditions including *Prosopis cineraria, Acacia tortilis and Lycium shawii*, together with salt-tolerant shrubs such as such as *Haloxylon salicornicum* and *Salsola* spp."

A second habitat restoration project that Kew and the MSBP are involved with in the Middle East is just getting under way. The hostilities that followed Iraq's invasion of Kuwait in 1990 left the nation in a sorry state, for which the United Nations agreed to pay compensation. There is much work to be done to restore and revegetate degraded land. Kew and the MSBP are working with Kuwait to train personnel and help them realise the Master Plan for Kuwait. "We will be working with them on areas they have demarcated as conservation areas," explains Shahina.

The sites stand on a gravel plain, the natural vegetation of which has been degraded by overgrazing and tourism. Fortunately, Kew's Herbarium resources include records made by Dame Violet Dickson, a botanist who lived in Kuwait for 61 years following World War I, and who published a flora of the country in 1955. The plant *Horwoodia dicksoniae* is named after her. Records also exist from explorer Sir Wilfred Thesiger's journeys around Southern Arabia in the mid 20th century, as well as duplicates of herbarium material made in Kuwait before the Gulf War.

In Kuwait two-thirds of the plants are annuals, which come up after the winter rains. The sparcity of vegetation make it susceptible to damage. "Kew staff will undertake a very detailed vegetation survey so we know what species are there already and which have disappeared," explains Shahina. "We will also look at all the literature, along with pre-war images. Meanwhile the MSBP will be advising on construction of a national seed bank and helping train seed collectors and data-recorders. And our horticultural team will be helping develop a nursery where plants to be used for the revegetation and restoration will be grown."

ABOVE: The Kew herbarium specimen sheet for *Horwoodia dicksoniae*.

OPPOSITE TOP: In the desert, recording data to inform restoration work.

OPPOSITE BOTTOM: *Acacia tortilis* growing in a degraded landscape in the United Arab Emirates.

Restoring habitats in Australia and the USA

MSB seed collections are underpinning habitat restoration work in Australia and the USA. The iconic Australian feather-leaved Banksia (*Banksia brownii*), named after Kew's 18th-century unofficial director Joseph Banks, only survives in the wild in three areas. Already hit hard by an introduced *Phytophthora* fungal disease and too-frequent forest fires, its numbers are set to dwindle further because of climate change. Expert Anne Cochrane, of Western Australia's Department of Environment and Conservation, fears it could be extinct within a decade without protective measures.

To prevent that scenario from becoming reality, in 2008, Anne brought seeds of the *Banksia* from Australia to the MSBP. Over ten weeks she worked with seed ecologist Matt Daws to discover how best to germinate the seeds. Most grew into seedlings, whereupon they were flown back to their homeland in an aeroplane hold in sealed plastic trays. Genetic studies show that the species comprises three distinct sub-populations: mountain, coastal and southern. Young plants were transplanted to sites most appropriate for maintaining local adaptations and

ABOVE: **Bison grazing in Konza Tallgrass Prairie in the Flint Hills of Kansas.**

genetic distinctiveness, where they are now helping to ensure *Banksia brownii's* survival in the wild.

Elsewhere in Australia, seed collections initiated by the MSBP's partnership with Seedquest New South Wales are helping save the Regent honeyeater (*Xanthomyza phrygia*). This endangered bird, of which between 1,000 and 1,500 mature individuals remain in the wild, occurs in open box-ironbark forests and riverside stands of *Casuarina* on the inland slopes of the Great Dividing Range, and in coastal forests in NSW and eastern Victoria. The bird needs access to nectar-giving plants such as *Eucalyptus* and mistletoe when breeding, and experts believe that clearance of important forest stands has led to the honeyeater's decline.

One of the honeyeater's last remaining strongholds is the Capertee Valley, located a three-hour drive northwest of Sydney. A revegetation programme has been under way here since 1993, with a range of locally collected plant species favoured by the honeyeaters used in the plantings. The most important of those species are Mugga ironbark (*Eucalyptus sideroxylon*), yellow box (*Eucalyptus melliodora*), white box (*Eucalyptus albens*) and river oak (*Casuarina cunninghamiana*), the latter because of its role as host to the mistletoe *Amyema cambagei*. The seed bank in NSW has provided 250 species of native plant seed every year to the MSBP, with the Capertee Valley a major focus of volunteers' collecting effort. The hope is that if the honeyeater's habitat can be preserved and restored, the bird, too, may avoid extinction.

In the USA, habitat restoration has become critical. All the MSBP's USA partners cited habitat restoration as the number one priority when considering how it would use seed collections and knowledge gathered

by partnering with the MSBP. 'The Seeds of Success Programme is delivering the wild seed collections that are the first step in the process required to restore native plant communities in the USA,' says Michael Way, International Coordinator for the Americas. Local partners have then taken the seeds and protocols gained through working with MSBP and put them to use in a wide range of restoration projects. For example, Chicago Botanic Garden has focused on the recovery of the central tall-grass prairie, which is one of the world's most threatened habitats, having been reduced to less than 0.01 per cent of its former range. Their role in the Seeds of Success Programme is to collect seed of priority species from the former prairie states of central USA, from Minnesota south to Oklahoma. In addition to the wide range of biologists using the collections, seed has been provided to a local native seed farm project and also to the University of Montana for grassland research.

MSBP partnerships have also contributed to the Great Basin Native Plant Selection and Increase Project. This began in 2001, with the aim of boosting stocks of native plants, especially forbs, and providing knowledge and technology so they can be used to restore diverse plant communities across the Great Basin. More than 20 federal, state, and private partners are involved in the project. 'Our partners have provided over 100 seed lots and contributed more than 100 common garden studies to the Great Basin Project, looking at the way a seed collection grown in one environment can be successfully moved and grown in another environment without failing,' explains Michael. 'The long-term aim is to reverse the former losses of habitat and prevent the invasion of alien species over this enormous area.'

Jie Cai, China

Interviews around the world

Jie Cai is one of several botanical experts who has one foot in the MSBP and one abroad. Now based at the Kunming Institute of Botany (KIB) of the Chinese Academy of Sciences (CAS) in Yunnan Province, he initially spent almost three years working at the MSBP in the UK. His role is that of Internatonal Collecting Coordinator, which involves coordinating seed collections in China along with international collaborative ventures.

'When the MSBP international programme started in 2000, the Chinese government was planning to set up a germplasm bank in 2004,' he explains. 'Because Kew is a leading organisation in plant science, and was at that time looking for international partners, the MSBP and CAS signed a ten-year agreement, the Wild Germplasm Conservation Agreement. There was a need for a person to liaise between the KIB and the MSBP, so I got the chance to become involved.'

KIB and the MSBP have the same goal to conserve global plant biodiversity. Jie Cai's work involves initiating and managing seed-collecting expeditions with various partners within China, working out which species to target for seed-collecting expeditions, gathering and analysing data on the collections held in KIB's seed bank and exploring opportunities with other potential international collaborators. So far, the KIB has banked seeds from 4,781 Chinese species. These include seeds from the limestone-loving plant *Paraisometrum mileense*. First recorded in 1906 by a French missionary, the plant was not seen again for 100 years. Only 320 plants are now believed to exist in the wild in Southwest China.

China has a rich and unique plant biodiversity. It contains some 31,000 plant species and is the only

ABOVE: **Jie Cai from the Kunming Institute of Botany, partner of the MSBP.**

AMAZING PLANT FACT

Sole survivor

Kew's links with China date back some two and a half centuries. The Gardens has a specimen of the Chinese maidenhair tree *Ginkgo biloba* that was planted in 1762, one of the first of the species to thrive in Britain following the tree's introduction to Europe in 1754. *Ginkgo biloba* is the sole survivor of an ancient group of trees that dates back to the time of the dinosaurs.

Seed cathedral

place on Earth where an unbroken connection remains between tropical, subtropical, temperate and boreal forests. Many plant species rendered extinct in North America by glaciers in the last ice age survive in China. The Dawn Redwood (*Metasequoia glyptostroboides*) is one example. This species was only discovered in 1949. The specimen at Kew is the oldest in cultivation, dating back to the first plantings between 1949 and 1951.

China's flora faces threats from urban developments, climate change and overharvesting for the trade in traditional Chinese medicine (TCM). The latter is now a global industry, but many of the plants used are still harvested from the wild. 'Quite a lot of these plants cannot yet be cultivated,' explains Jie Cai. 'We are undertaking some early-stage work to try and cultivate TCM species. For example, we are trying to help local farmers in southeast Yunnan to grow *Panax pseudoginseng*.' This is an important herb used in Chinese medicine to improve the overall health of the human body; in Latin, Panax means 'cure-all'.

The Kunming Institute of Botany is government-funded, with some 300 staff and almost as many doctoral and masters students. Its facilities include a centre for advanced scientific and technical information, a centre for apparatus analysis and tests, herbarium, seed bank and botanic gardens. It concentrates on four research fields: plant evolutionary biology, phytochemistry and chemical biology, plant genomics and conservation biology. The MSBP and KIB collaborate on several programmes, including staff exchanges, seed exchanges and in co-supervising PhD students.

Seeds from the Kunming Institute of Botany were used to highlight the world's biodiversity within the British Pavilion at the 2010 Shanghai Expo in China. The brainchild of designer Thomas Heatherwick, the Pavilion comprised a 15-metre-high wooden cube, pierced with 60,000 7.5-metre-long clear acrylic rods, giving it the appearance of a 'hairy', textured building. The rounded, interior ends of the four-sided batons contained seeds that shined in earthy hues as daylight passed through from the outside. Unlike other Pavilions at the Expo, the British entry had no flashing lights or noisy screens reflecting the digital age. The idea was that visitors should wander in the quiet calm of the cathedral-like space and contemplate the astonishing biodiversity of the plants we rely on for food, clothing and medicines.

ABOVE: Four images of the 'Seed Cathedral', the UK's pavilion, at the 2010 World Expo in Shanghai, showing the display created by more than 60,000 acrylic rods, each containing a seed sample (*see also page 174–175*).

Masego Kruger, Botswana

Interviews around the world

Just as Jie Cai forms a link between seed conservation efforts in the UK and China, so Masego Kruger-Gaadingwe connects the MSBP with Botswana. Masego works as a Senior Research Officer at Veld Products Research and Development, an NGO specialising in native plant products. Based in the village of Gabane, close to Gabarone, the NGO aims to use Africa's natural resources in a sustainable way. It conducts fruit tree planting and agroforestry trials, as well as undertaking community-based projects. Fruit trees are important in Botswana, as they can yield a crop in poor-rainfall years when arable crops fail.

Masego joined Veld Products in 2003, after taking an MPhil in Applied Ecology. Her role, which is funded by the MSBP, has entailed identifying more than 40 priority species in Botswana, locating populations of those plants, gathering seeds from them and growing seedlings in order to draw up propagation protocols that can be passed on to local communities. 'The main aim of our project was to prioritise at least 40 useful plants from the country and then conduct propagation trials,' she explains. 'They were selected for their usefulness to the community. We looked at which parts of the plants were being used, and whether harvesting those parts was being done sustainably. Based on that we decided whether species were vulnerable or not.'

Each day, Masego and her assistants monitor the progress of plants in the NGO's greenhouse and shade nets, recording data on which plants have germinated or come into flower. She is also regularly involved with seed-collecting expeditions with national partners across Botswana. Botswana has around 2,300 species, which grow in habitats ranging from the parched Kalahari Desert to the wetter north, where there are plentiful rivers and taller trees. Rapid development of the country in recent years has put pressure on Botswana's flora. There are 43 species on the country's Red Data list, of which 13 are considered critically endangered, endangered or vulnerable and 22 are data deficient.

One of the species that Masego's team has successfully collected and propagated is *Stomatostemma monteiroae*. Known colloquially as 'tree meat', this plant is important to rural communities in Botswana as a meat substitute. 'This species had never been propagated before in Botswana,' explains Masego. 'Communities have been simply harvesting it from the wild without replenishing the stocks. We found out that it wasn't very difficult to propagate and we successfully produced seedlings.'

Once propagation protocols are available, Masego and her team go into the communities to pass on the information to people who use the plant. They also donate seedlings grown up in the trials by the project. For example, some tree meat plants have now been donated to communities in Tswapong district, which sell the fruits of the plant. 'On occasions such as National Tree Planting day, we go and attend shows to demonstrate to people that they can grow some of the plants they usually harvest from the wild and to give them seedlings. We aim to show people that the plants they eat or use for other purposes can be grown in their backyards.'

Masego has also been involved with reintroducing plants in the north of Botswana. Three of the plants earmarked as priority species by the project, *Berchemia discolor*, *Euclea divinorum* and *Hyphaene petersiana*, are used by northern communities to make baskets. 'It was known as far back as 1986 that a lot of plants around the villages had disappeared and that people were

having to travel farther to gather material to make baskets. We donated some seedlings for reintroduction and I went there with some of the community people to show them how to plant the seedlings and grow seeds for themselves.

'The great thing about my role is that I am constantly learning. I know some of the plants I work with because I grew up with them but out there in the wild I've come across things with features and characteristics that I've never seen before. I also enjoy working with the various communities because I get to learn about the importance of different species. As researchers we sometimes just consider plants in terms of collecting seeds and preserving biodiversity, but in the communities you get a whole new perspective. Its interesting to see how the indigenous knowledge that communities have about the plants they use is often backed up by the scientific work that we do.'

AMAZING PLANT FACT
Slimming pressures

Hoodia currorii, one of Botswana's native species, is used by San bushmen to stave off hunger pangs. Meanwhile, its close relative, *Hoodia gordonii*, has been found to contain a compound that can suppress appetites. Both species are now subject to intense harvesting in the wild, as companies believe their active compounds could form the basis for a slimming pill. During an expedition to Botswana in 2003, the MSBP collected seeds from four separate populations of *Hoodia currorii* to safeguard the species for the future.

WILLOWMORE CEDAR

SOUTH AFRICA: About 120 km west of Port Elizabeth the Baviaanskloof Wilderness in the Eastern Cape Province is 270,000 hectares of unspoiled, rugged mountainous country.

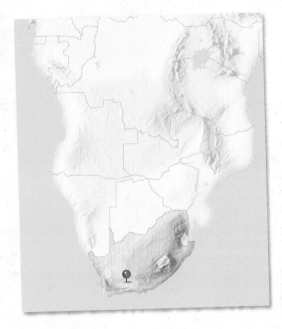

Wood from this tree is remarkably durable and resists decay. Early settlers in the area use the yellow aromatic wood for most of their timber work. They build houses, and make fences, telegraph posts and furniture.

ABOVE: The challenging terrain that is home to the Willowmore cedar.

ABOVE: A magnificent cedar in its natural habitat.

The tree is named after two people: a late 18th-century Royal Naval captain and a 20th-century cactus collector: Edward Widdrington (sailor and conifer botanist) and Friedrich Schwarz.

This is wild terrain, home to a variety of animals including the Cape leopard, Cape buffalo, eland and baboon. But the team of plant hunters from the MSBP in Cape Town is determined to find and collect seed from a rare endemic, an ancient giant of a cedar tree known as *Widdringtonia schwarzii*, the Willowmore cedar.

ABOVE: The Baviaanskloof Wilderness in the Eastern Cape.

Some trees are within scrambling reach and they manage to collect a few cones, but the best cone-bearing trees are much higher up, way out of reach. To get to these more mature trees, rope and a head for heights are required, as the only way to reach them is to climb above the ravine and then abseil down the cliff face.

Widdringtonia schwarzii grows at heights above 70 metres but no higher than 1,220 metres, so finding specimens can involve a little light mountaineering. This expedition is no exception. The team leave the vehicle well before they reach the area in which they hope to collect seed because they don't want to damage any vegetation. Equipped for climbing, they set off on foot to climb in search of the trees.

They make a good collection and climb back to the top of the cliff, then hike back to the vehicle. It's been an exhausting but immensely successful trip and the black-brown, broadly-winged seeds of *Widdringtonia schwarzii* are safely deposited, half in the MSB in the UK and half in South Africa's National Plant Genetic Resources Centre near Pretoria.

ABOVE: *Widdringtonia schwarzii* close-up.

ABOVE: **Exploring the wilderness.**

After walking for some time they have to scramble over rocks for about an hour but eventually spot some trees with gnarled trunks of red-grey bark nestling in a ravine. These are what they've come so far to find: the Willowmore cedars.

The team climbs higher, ropes up and makes the tricky descent to the trees. The seed collectors are looking for female cones, which are dark brown, with rough, warty scales. Cones in various stages of development are on the tree all year round.

Plant profile

COMMON NAME: Willowmore cedar

LATIN NAME: Widdringtonia schwarzii

FAMILY: Cupressaceae

STATUS: Red Data Listed (Near Threatened 2006)

SIZE: Can reach 40 metres

DESCRIPTION: Trees often have gnarled trunks and spreading branches. Bark is red-grey, thin and fibrous. Leaves are grey-blue and ovate. Juvenile leaves are needle-like. Male and female cones are borne on the same plant.

THREATS: Runaway fires, off-road tracks.

Propagating threatened cacti and bulbs in Chile

BELOW LEFT: *Echinopsis chiloensis* ssp. *littoralis.*
BELOW RIGHT: *Menodora linoides.*
OPPOSITE , CLOCKWISE FROM
TOP LEFT: *Eriosyce subglobosa* var. *clavata,* Menodora linoides; *Eriosyce subglobosa* var. *clavata, Echinopsis chiloensis* ssp. *littoralis.*

Chilean botanists are working with horticultural experts and botanists at Kew to find ways of propagating some of Chile's most threatened plants, including desert cacti and bulbs such as rare kinds of amaryllis. Many species that grow in Chilean Mediterranean areas and the desert regions are threatened by human activity, especially habitat loss, as agriculture and construction move into areas that used to be comparatively untouched by humans.

The hope is that seeds harvested from plants propagated by the project (The Ex situ Conservation of Threatened Chilean Flora Through Propagation) will be used to bolster populations of rare species in the wild. Conservationists in Chile are working to propagate these species in ways that will maintain their natural genetic diversity. Traditional crop breeding tends to narrow the gene pool and leads to uniformity, rather than diversity. Such ecologically minded propagation schemes may actively boost conservation of Chile's rich native flora, as well as promoting seed banking and the protection of natural habitats.

So far, a total of 15 priority species have been successfully propagated, either by seed or vegetatively, including *Dalea azurea*, a critically endangered species in the legume family, and *Placea lutea*, a rare member of the Amaryllidaceae. *Dalea azurea* is now restricted to one valley in Paposo, in the Atacama region in Chile. Plants survive in the wild in such low numbers that only a very few seeds could be collected in any one season. Now that successful *ex situ* propagation means that *D. azurea* can be grown in the greenhouse, the next generation of seeds can be harvested and conserved in seed banks in Chile and Wakehurst, ensuring the long-term conservation of the species. Meanwhile, some of the greenhouse-propagated plants can be reintroduced into the wild in the future.

The MSBP helped to convene an international workshop, which took place in Chile in 2008, to share the most effective techniques for propagating wild species. With financial support from the mining company Rio Tinto plc, Kew staff shared their experience in the field, laboratory and greenhouse. Seventy conservationists from Peru, Argentina, Ecuador and Chile attended the workshop, while 30 specialists stayed on for two more days of intensive training on different plant groups. The participants learnt new skills to use in nurseries, universities and botanic gardens.

Small nurseries across Latin America are now being set up to propagate native floras for restoration schemes and to protect wild plants from over-exploitation. Kew's involvement has helped to raise the profile of the project, share skills and boost enthusiasm for working with native species. There's now a real fascination in Chile for growing these often-threatened plants.

Rosemary Newton

Seed Germination Specialist

OPPOSITE: **Rosemary Newton working on her PhD at Wakehurst Place.**

Rosemary Newton spends her days trying to persuade troublesome seeds to germinate. All seeds that enter the MSBP vault undergo germination tests to make sure they are able to produce plants if required in future. Those that stubbornly refuse to put out a shoot become Rosemary's wards. She investigates the ecosystems they inhabit in the wild and tries to coax them to life by duplicating those environmental conditions in the laboratory. 'I love the challenge of having a problem and solving it,' she says. 'When you have something you haven't been able to germinate and then you get good germination that's really thrilling. It's incredible to think that a plant can produce a little capsule that will survive until conditions are right and then produce another plant.'

Born and brought up in South Africa, Rosemary obtained her undergraduate degree at the University of the Witwatersrand in Johannesburg. She later moved to Cape Town where she studied botany for a Masters degree, specialising in the germination of the nut-fruited Restionaceae, one of three main plant families of South Africa's Cape Floral Kingdom. 'I found the dormancy puzzle in one particular species, *Cannomois virgata*, a particularly tough one to solve,' she explains. 'Seeds from these plants need a period of soil storage that subjects them to fluctuating temperatures and moisture to break their dormancy. They then require a smoke cue to signal that the surrounding vegetation has been cleared, the competition has gone and it is time to germinate. I spent 18 months testing different dormancy-breaking treatments and was finally successful on the very last experiment.'

The work sparked Rosemary's love affair with seeds, which brought her eventually to the Millennium Seed Bank in 2005. With every new batch of seeds that arrives on her desk, she begins an investigation to try and work out what will make them spring to life. This involves initially searching existing literature, the MSBP's Seed Information Database and the internet to see what information she can glean on the plant's native habitat and any past germination attempts. 'I'll then try and design an experiment that will interpret the data,' she says. 'For example, if the information suggests the seed has a physical dormancy, I'll chip the coats of some seeds and leave others unchipped. If there is a physical dormancy, the chipped ones will germinate and the unchipped ones won't.'

The tools of Rosemary's trade are: incubators that can be programmed to deliver different temperature and light levels; agar jelly to provide a moist base for the seeds to rest on; and chemicals such as gibberellic acid and potassium nitrate. These can help speed up germination; for example, gibberellic acid tricks seeds into thinking they have undergone a cold spell, an event that is vital for some seeds to start growing. 'I'll try different temperature and light regimes,' says Rosemary. 'And if the plant comes from a fire-driven ecosystem, then I'll consider adding a heat or smoke treatment. If the plant lives in a hot desert, I won't subject the seeds to a long, cold spell, but if it comes from Europe, then I might. I'll also look at whether in nature, seeds are shed into a wet, dry, warm or cold season.'

The processes that seeds undergo before they germinate can be highly complex. For example, seeds that come from areas of the world with a notable dry season have to delay germinating until the rainy season, even if there are sometimes short bursts of rain during the dry season. A complicated process

'It's incredible to think that a plant can produce a little capsule that will survive until conditions are right and then produce another plant.'

of chemical changes within the seeds enables them to hold off germinating until conditions are optimal. 'If seeds are shed at the beginning of the dry season they'll have a physiological block that prevents them from germinating, even if they are exposed to water. As the warm season progresses, this will slowly break down. It's incredible how these species have evolved to allow them to survive prolonged unfavourable periods and then germinate in the correct season,' she says.

For the past three years Rosemary has juggled her daily detective work on the MSBP collections with studying for a PhD on dormancy in the European Amaryllids, a family that includes snowdrops (*Galanthus*) and daffodils (*Narcissus*). Her studies have revealed that summer is very important for these species, as the warm months allow the embryos to grow and physiological dormancy to be broken. The arrival of cooler autumn days then triggers germination. 'If you put these seeds directly into the soil in autumn they won't germinate, as they need the summer warmth,' says Rosemary. She has also found that drying the seeds affects their longevity, something that has implications for their storage in the MSBP vault, as drying is a necessary part of the storage process. 'The trouble is, the more we find out about seeds, the more questions there are to answer,' she sighs.

Chapter 6

Breathing life into degraded ecosystems

The magnificent *Seed Cathedral*, designed by Thomas Heatherwick,

Saved seeds spring to life and restore damaged habitats

ABOVE: **Bloomers Valley grassland.**

OPPOSITE LEFT: **The rich flora of the central grassland.**

RIGHT: **Iain Parkinson getting his hands dirty planting out wildflower seedlings.**

Not far from the MSB building at Wakehurst Place, past the national birch collection where primroses and bluebells colour the ground in spring, lies Bloomers Valley. Fringed by woodlands, the central grassland is the site of the MSBP's first major habitat restoration project. The aim is to reintroduce around 30 species of native plants and turn the hillside back into a species-rich meadow. Restoring damaged habitats is becoming increasingly important in light of the drastic biodiversity loss affecting many parts of the globe. In future, such 'ecological restoration' is likely to become a major use for the MSBP's seeds and the study of it, or 'restoration ecology', a major focus of Kew's research.

'At Wakehurst, we haven't used seeds for a large-scale restoration project until now so we're testing the water,' explains Kate Hardwick, Restoration Ecology Coordinator. 'We're doing it partly to gather data and partly to find out the best way to conduct such experiments; to find out what sort of things we should be testing and setting controls for, and discover what problems are likely to arise. We had a meeting with our MSBP partners and asked them all what new help they wanted from the seed bank and loads of them mentioned restoration. They want to be able to use the seeds they collect.'

The three-year restoration programme, funded by the John Ellerman Foundation, began in autumn 2009. The MSBP didn't have sufficient seed stocks for planting an entire meadow, so Wakehurst's Conservation and Woodlands Manager, Iain Parkinson, acquired seeds from a local NGO. They chose species that are characteristic of neutral and acidic grasslands and also looked at the specific conditions of the site, which is damp in some places and drier in others. With grass already present, the plants chosen were mostly herbaceous perennials such as common knapweed, autumn hawkbit and musk mallow.

The first hurdle came when the team tried to germinate the seeds. Collections of most of the species were banked at the MSBP, so germination protocols were available. However, despite following those instructions, around half of the species failed to germinate. A couple germinated very well and the results for the rest ranged from acceptable to poor. Thinking the seeds might be immature, MSBP staff weighed a selection of individual seeds and compared them to those in storage. They found they were, if anything, bigger than those in the bank, so that was not likely to be causing the problem.

A visit to see the seed storage facilities revealed what the trouble was. It turned out that some of the seed had been collected the previous year and stored in fabric sacks on a barn floor. This may sound harmless enough, but the fluctuations in temperature and humidity throughout the year had been enough to kill the seeds. Because the MSBP follows strict seed handling and storage protocols that ensure high seed quality, it hadn't even occurred to the team that viability might be a problem. 'We took it for granted we were getting viable seeds,' Kate says. 'But our storage protocols weren't obvious to our partners. It was a real reality check for us.' The MSBP is now working closely with its UK collaborators to advise on seed-storage issues so that in the future this valuable local product will be of the highest quality.

Undeterred by this set back, Kate, Rosemary Newton (the MSBP's Seed Germination Specialist) and their team planted more seed to ensure they had sufficient seedlings to plant out. They experimented with three

different methods of germinating seeds, so they could gather data about the most effective methods for this kind of restoration. 'For the low-input treatment, we sowed the seed directly on the ground,' explains Kate. 'We scarified and harrowed the ground until it was 40 per cent bare earth and then sowed the seed evenly across the site. We created 20 monitoring quadrats and 20 covered 'exclusion' quadrats to assess the results. Using an alternative method to test this, we also sowed 100 seeds of each species in individual one m² plots.'

The moderate-input treatment involved sowing seeds in pots in the nursery and transplanting them into modular trays for planting out in the meadow, while the high-input method required seeds to be sown on agar plates in highly controlled laboratory conditions using incubators set at known temperatures to germinate the seeds. Half the seedlings grown up using these methods were then transferred into compost and half into soil taken from the site. 'We did this to see if the site soil would give an advantage; we thought the seedlings might be hardier or there might be mycorrhizal fungi in the soil that would benefit the plants,' explains Kate.

On a warm spring morning, Kate inspects her seedlings to assess progress. In the nursery, the rows of tufted vetch and common knapweed planted in compost seem to be flourishing, while those in the site soil appear rather stunted by comparison. Out in the meadow, meanwhile, there are early signs that some of the scattered seeds have taken root. 'These curly looking leaves are of yellow rattle,' Kate explains, pointing to a diminutive seedling as she scrutinises the plots on hands and knees. 'That's a good sign as they are a parasitic plant that reduces the vigour of the grass, enabling other species to become established.'

RIGHT: **Two views of**
Madagascar: degraded land
contrasts with abundant
natural beauty.

'Species rich grassland is a precious habitat that's rapidly disappearing,' adds Kate. 'The restoration and active management of grasslands like Bloomers Valley will increase their wildlife value and maintain locally diverse landscapes for future generations to enjoy.'

Although it is early days for the MSBP's own restoration project, the field of restoration ecology has been developing since the 1940s. An early visionary in the USA was the conservationist, forester and philosopher Aldo Leopold, who believed that humans should view the natural world 'as a community to which we belong'. Closer to home, the former Chair of Botany at Liverpool University, Tony Bradshaw, was one of the first people in the UK to recognise the value of restoration ecology. His work on revegetating china clay tips in Cornwall formed the basis of the techniques used to develop the Eden Project site.

Kate Hardwick's interest in restoration ecology dates back to when she wrote her BSc thesis on the restoration of spoil from the Channel Tunnel. When she followed up this work by studying for a PhD on restoring tropical rainforests in Thailand, the benefits of restoration ecology were still little understood. 'While I was working for an NGO trying to replant native forests in the early 1990s, the Thai Forestry Department was planting non-indigenous pine and Eucalyptus trees and thought what we were trying to do was ludicrous,' Kate explains. 'People called it 'jungle gardening'. Things have changed massively since then; now everyone's interested in it.'

Today, restoration ecology is better known but still evolving. While restoring a meadow might be relatively straightforward, returning other ecosystems to their native state can be far more complex. Recent research has revolved around 'thresholds' and 'stable states'.

'If you degrade an ecosystem a little it can bounce back,' explains Kate. 'However, if you pass a certain threshold, other factors come in that stop it returning to where it was before. For example, where I worked in Thailand, if you cut down a few trees, new trees would grow very quickly. But if you cut down a lot of trees, grass would start to invade.'

In the tropics, the trouble when grasslands establish themselves in former forest is that wildfires are more likely. Once fire affects an area, young seedlings burn so the trees are no longer able to re-establish themselves. Birds and mammals that dwell in the forest canopy no longer have a home and so move away; in doing so, their important seed dispersal role is diminished. 'You have to find a way to take the ecosystem back past the threshold to a point where trees can establish themselves once more,' says Kate. 'One way of doing this is to introduce 'framework species' such as fast-

growing trees. These act as a catalyst to shade out the grass and attract animals back.'

Understanding the suites of plants that help form ecosystems is where Kew comes in. So far the Gardens have contributed knowledge or resources to 60 or so external habitat restoration projects. A number of major programmes are ongoing in the United Arab Emirates and other parts of the Middle East, where government authorities are restoring large areas of desert degraded due to overgrazing, recreational use and past warfare. Shahina Ghazanfar, Head of the Temperate and Middle East Team in Kew's herbarium, is helping reconstruct desert habitats by surveying existing vegetation to assess the baseline flora, consulting records at Kew and using examples of existing undisturbed ecosystems.

Meanwhile, Kew's Geographic Information System (GIS) Team are creating restoration maps that use geographical layers, such as topography and climate, to predict how habitats can best be restored across large tracts of desert. Teams from the MSBP and Kew's propagation units are also working together to create restoration facilities at sites across the Middle East. 'These are good examples of projects where we're using many of Kew's skills and working together,' explains Kate.

'In any restoration project you have to identify a reference habitat that represents the ecosystem you are trying to get back to. Kew has taxonomists who can identify plants, along with historic records in the Herbarium that show which plants once existed in now degraded sites. After that you need plant material; we have seed collection and germination expertise, as well as skills in setting up seed banks. Once you have germinated the seeds, you have to plant them out, which requires horticultural expertise. The skills and resources that Kew has are all relevant to restoration ecology.'

Helping partners use 'difficult' seeds

If farmers are to cultivate particular plants instead of exploiting wild supplies, they must know how to germinate seeds. It sounds simple, but many seeds need specific environmental conditions to coax them to break dormancy and then germinate. Others might be easy to germinate but require careful storage to ensure their long-term viability. MSBP scientists use an array of incubators simulating natural temperature and moisture regimes to crack germination codes of new seed arrivals so they know they are viable and acceptable for *ex situ* banking. By ensuring this knowledge reaches

communities that rely directly on those plants for food, medicine and building materials, they are also helping conserve wild plant populations *in situ*.

The 'Difficult' Seeds project illustrates that knowing how to germinate seeds can support farmer's efforts to diversify their livelihoods as well as aiding conservation. This ongoing collaboration between Kew and the Food and Agriculture Organization of the United Nations (FAO) is helping seed banks and farmers better identify, handle, store and use seeds of problematic species.

At its outset in 2006, staff from seed banks in 29 African

ABOVE & RIGHT: Difficult
Seeds' Workshops in
Morocco and Burkina Faso.
OPPOSITE: Seeds of *Euclea
undulata.*

countries identified 200 'difficult seeds'. 'This picked up species that were tricky to store, whose dormancy characteristics make them difficult to germinate and those where handling issues meant seeds were losing viability quicker than they ought,' says Kate Gold, Training Manager.

The species in question ranged from shea, *Vitellaria paradoxa*, whose seeds are easy to germinate but hard to store, to the leafy vegetable *Cleome gynandra* which requires chipping at both ends of its seeds for successful germination and the bottle gourd *Lagenaria siceraria*, whose seeds must be fully ripe before extraction to be useable (see box). In some cases, seeds of maize and sunflower, for example, were becoming unviable simply because they were not being dried sufficiently before storage. 'If sophisticated drying facilities are not available, it's easy to use a desiccant such as charcoal,' explains Kate. 'This is widely available in Africa and can be dried in the sun. When placed in a plastic bucket with a lid on, it creates the perfect environment to dry seeds.'

MSBP staff Moctar Sacande, Kate Gold and Vanessa Sutcliffe ran workshops in Kenya, Burkina Faso, Botswana and Morocco. These were attended by 60 participants from 48 institutes in 38 countries, with 80 farmers in the host countries attending associated farmer's sessions. In Morocco they helped farmers who were having trouble keeping their grain seeds viable from one season to the next. They stored grain in concrete underground bunkers called *mat moura*. Technology Specialist Vanessa Bertenshaw and her colleagues put hygrometers and thermometers in the bunkers, which revealed that they were very hot and humid, the worst conditions for keeping seeds alive. 'We worked with the farmers to find solutions,' she

Floating containers

The bottle gourd (*Lagenaria siceraria*) is one of the earliest plants cultivated by humans. An annual herb considered by most experts to be native to Africa, it was used on that continent as far back as 5,000 years ago for food, medicine and as a source of utensils. Archaeologists believe it reached temperate and tropical areas in Asia and the Americas even earlier. It may have made this journey with human help or more likely as a wild species whose fruits had floated across the seas. Its fruits can float in the sea for many months without seeds losing their viability.

says. Sealing the walls of the bunker to stop moisture seeping in, then using charcoal to dry both bunker and seeds seems to have made a big difference. Farmers also got useful tips from each other. One farmer from Sudan got rid of insects by treating the seeds with smoke from neem leaves. 'The Kenyan farmers hadn't thought of that and were keen to try it. It was a really nice exchange of ideas. The workshops revealed that many difficulties experienced by genebanks and farmers trying to germinate seeds are solvable with existing information. With this in mind, the MSBP is developing web pages on seed biology and tried-and-tested germination and storage methods for each of the 'difficult' species. With climate change set to make some crops less viable and others more suited to a warmer climate, being able to learn from the MSBP's germination experiments could make all the difference for an African subsistence farmer between cultivating a successful crop and having to fall back on dwindling wild stocks for food.

Seeds of *Sterculia setigera*, a species from tropical West Africa.

Founding heroes of the Millennium Seed Bank Project

JACK HESLOP-HARRISON

Following the first United Nations 'Earth Summit' in Sweden in the 1970s, the then Kew Director Jack Heslop-Harrison established Kew's first facility for storing wild seeds, with the aim of examining if wild seeds could be stored with the same ease as crops seeds. In the *Journal of the Kew Guild* in 1972 he wrote: 'It is impossible to believe that we will reach the end of the century without the national recognition of the need to hold our own substantial stock of important genera, and contribute this to conserving their worldwide variability as human pressure on the plant kingdom increases.'

PETER THOMPSON

Peter Thompson joined Kew in 1964 as the organisation's first plant physiologist, choosing to focus on seeds. In his memoirs he recalls botanical gardens' existing seed exchange programmes: 'Throughout the world seeds were being collected, cleaned and put into packets; lists were distributed; time was spent on trying to grow them, and much of the effort was wasted because a high proportion of these collections was wrongly named, dead or even non-existent.' He recommended installation of a seed-handling and testing laboratory and instigated the first external collections for the bank, of seeds from the Mediterranean flora. The Physiology Section incorporating the seed unit moved to Wakehurst Place in 1974.

SIR GHILLEAN PRANCE

Sir Ghillean Prance joined Kew as Director in 1988. He was a passionate environmentalist and was responsible for expanding Kew's role in conservation. At a seminar for Kew staff in 1993, he expressed his wish for a 'large modern underground seed facility', and outlined bold objectives for the year 2020 that included the Seed Bank having collected 20 per cent of the world's flora. In 1995 he received a knighthood for his services to conservation. He retired in 1999 just before the Wellcome Trust Millennium Seed Bank opened.

MARGARET THATCHER

The former prime minister might seem an unlikely hero for the MSBP but her reforms changed the Royal Botanic Gardens, Kew from being a part of the civil service to a Non-Departmental Public Body (NDPB) reporting to a Board of Trustees. She also introduced 'science audits' for such NDPBs. Kew's first audit in 1990 questioned the level of impact the seed-banking facility at Kew was having, ultimately spurring on its management team to develop a more ambitious project. They were rewarded for their efforts when the Millennium Commission awarded them £30 million in 1995. A stipulation of the award was that either Kew or external sponsors had to match this funding.

GILES COODE-ADAMS

The transition to a NDPB allowed Kew to establish a Foundation to raise funds from non-government sources. Giles, an investment banker with a passion for plants and concern for the planet, took on the role of Chief Executive of the Foundation. Following the launch of the Millennium Seed Bank Appeal in May 1996 by HRH Prince Charles, with Sir David Attenborough as patron, Giles secured substantial sponsorship including donations from premier sponsor Orange plc and the Wellcome Trust.

ROGER SMITH

In 1974 Roger Smith joined Kew, working in the newly-established seed unit at Wakehurst Place under Peter Thompson, whom he succeeded in 1980. A key driver of the Millennium Seed Bank Project, he developed the concept for the project and led it through to fruition, with the opening of the Wellcome Trust Millennium Building in 2000. As leader of the Project, he directed the gathering of seeds from the UK's flora, completed the same year, and also initiated the second part of the project involving international partners. Roger was awarded an OBE in the Queen's New Year Honours in 2000, for services to the Project. He retired as Head of the Seed Conservation Department and Leader of the MSBP in 2005 to undertake a fundraising role. Paul Smith succeeded him (*see job profile on Paul Smith on page 48*).

Saving the endangered triangular club rush

Visitors to the beautiful Tamar Valley of south-west England would probably walk past the triangular club rush without giving it much thought. With clusters of lime green stems reaching 150 cm tall, and fairly non-descript brown flowers, it's not one of the lookers of the plant world. However, those who spy it in the River Tamar on the Devon-Cornwall boundary would do well to give it a second look. One of Britain's rarest plants, it remains only in a handful of locations along this waterway and its survival hangs in the balance. If it weren't for the combined efforts of the MSBP, Kew Gardens, the Environment Agency (EA), Natural England and Panscape Environmental Consultants, it might already have become extinct.

Experts realised in the mid 1990s that the triangular club rush (*Schoenoplectus triqueter*) was in trouble. It had once grown in the Arun, Medway and Thames rivers, as well as the Tamar, but over the years had dwindled to a few isolated populations. As a priority species under the UK's biodiversity action plan, it fell to the Environment Agency to find a way to safeguard the species. 'At that time, we contacted Kew and took some plants from the Tamar for propagation by the Gardens' experts,' explains Jess Thomasson, Biodiversity Officer for the EA. 'Kew took seeds from those plants for the MSB and gave us lots of seedlings to reintroduce to new sites in the Tamar.'

A survey of the Tamar conducted in 2009 revealed that, despite these efforts, the plant had almost disappeared from the original site; only the transplanted seedlings remained. A new attempt at saving it was needed. The MSBP agreed to donate seeds, which were grown into 280 seedlings at Wakehurst Place. Those plants were brought back to the Tamar during the summer of 2010. Some were planted in locations that had proved favourable in the past; the others were used to create a nursery of plants so EA staff had extra stocks for transplanting in future. The National Trust allowed the nursery to be created in the grounds of the medieval house, Cotehele, near Plymouth. This meant its wardens would be at hand to keep an eye on the plants.

No one knows why the plant has struggled to survive in the Tamar and become extinct in other estuaries. It could possibly be due to pollution, climate change zor simply the unstable nature of its chosen habitat. In the UK it is living at its most northerly extent in Europe. It favours the brackish environment of estuaries and it generally grows at the top of the inter-tidal area, where mud is unstable and prone to slippage. 'The nursery is on a tributary of the Tamar called the Morden stream,' says Jess. 'It's a very sheltered environment. We've created a shallow hollow on a small island, which has a channel from the Morden stream so it will get some saline water at times. We've tried to mimic its natural environment but in a very sheltered place. 'We felt something had to be done or the remaining populations would become extinct.'

The EA and partners are hopeful that the new nursery will help safeguard the species and, in time, enable them to transplant the triangular club rush to several new locations. Some potentially suitable spots have been identified, all within a 7 km reach of the upper Tamar estuary. Since the last life-saving attempt more than a decade back, the EA's understanding of the sedge's ecology and habitat requirements has increased, and Jess is now more confident of saving the rare plant for posterity. 'It's our last chance to save this critically endangered species,' she says.

ABOVE: Taking club rushes to the re-introduction site.

OPPOSITE: Joanna Walmisley preparing to plant one of the club rushes.

Paying the price for stealing plants

The spiked rampion, which has showy spikes of white flowers, survives in only a few isolated populations in the UK. Legendary character Rapunzel is believed to have been locked up in a tower for stealing white rampion plants; hence the plant is named white Rapunzel on the continent. Protected by a Biodiversity Action Plan, anyone stealing it today would also be punished – but more likely by a fine than by isolated imprisonment.

Find out more

SELECTED BOOK PUBLICATIONS

Desmond, R. (2007). *The History of the Royal Botanic Gardens, Kew, 2nd edn.* Royal Botanic Gardens, Kew.

Flanagan, M. and Kirkham, T. (2009). *Wilson's China: A century on.* Royal Botanic Gardens, Kew.

Fraser, M. and Fraser, L. (2011). *The Smallest Kingdom: Plants and plant collectors at the Cape of Good Hope.* Royal Botanic Gardens, Kew.

Fry, C. (2006). *The World of Kew.* BBC Books, London.

Fry, C. (2009). *The Plant Hunters: Adventures of the world's greatest botanical explorers.* Andre Deutsch, London.

Heywood, V.H., Brummitt, R.K., Culham, A., Seberg, O. (2011). *Flowering Plants: A concise pictorial guide.* Royal Botanic Gardens, Kew and Firefly Books, Toronto.

Paterson, A. (2009). *The Gardens at Kew.* Frances Lincoln Ltd, London.

Smith, R. *et al.* (2004). *Seed Conservation: Turning science into practice.* Royal Botanic Gardens, Kew.

Stuppy, W. and Kesseler R. (2008). *Fruit: Edible, inedible, incredible.* Papadakis Publishers, London.

Stuppy, W. and Kesseler R. (2006). *Seeds: Time capsules of life.* Papadakis Publishers, London.

Wood, C. and Habgood, N. (eds) (2010). *Why People Need Plants.* Royal Botanic Gardens, Kew.

REPORTS AND SERIAL PUBLICATIONS

Kew. Quarterly magazine from the Royal Botanic Gardens, Kew.

Plant Diversity and Climate Change: A review of Kew's science activities relevant to climate change. (2010) Royal Botanic Gardens, Kew, UK. Online at – www.kew.org/ucm/groups/public/documents/document/kppcont_028817.pdf

Plants Under Pressure in a Global Environment: The first report of the IUCN Sampled Red List Index for Plants. (2010) Royal Botanic Gardens, Kew, UK. Online at – www.kew.org/ucm/groups/public/documents/document/kppcont_027709.pdf

WEBSITES

Biodiversity International – www.biodiversityinternational.org

Department of Environment, Food and Rural Affairs, UK (Defra) – ww2.defra.gov.uk

Convention on International Trade in Endangered Species of Wild Flora and Fauna (CITES) – www.cites.org/

Food and Agriculture Organisation of the United Nations (FAO) – www.fao.org

Global Strategy for Plant Conservation (GSPC) – www.cbd.int/gspc/

International Union for Conservation of Nature (IUCN) – www.iucn.org/

Kew's Breathing Planet Programme: Science and conservation at Kew – www.kew.org/science-conservation/index.htm

Kew's Millennium Seed Bank Partnership – www.keworg/science-conservation/save-seed-prosper/index.htm

Millennium Ecosystem Assessment - www.maweb.org/en/index.aspx

National Center for Genetic Resources Preservation – www.ars.usda.gov/Main/site_main.htm?modecode=54-02-05-00

Royal Botanic Gardens, Kew – www.kew.org

Seeds of Success – www.nps.gov/plants/sos/

Kew's Millennium Seed Bank and the future

This book celebrates the flagship international plant conservation partnership of the Royal Botanic Gardens, Kew. In 2009, Kew reached its 250th anniversary as the world's leading botanic garden and science powerhouse for plant diversity. Like all great projects undertaken by Kew and its partners over this extraordinary period of continued contribution and global relevance, the Millennium Seed Bank Project had a long gestation, a carefully planned and executed birth, and has been designed to last.

The book highlights the Project's history under the special guidance of four of Kew's Directors. Professor Jack Heslop Harrison's visionary directorship in the early 1970s firmly set Kew on the path of helping conserve the Earth's plant life. He identified the need for a national and international seed bank, and put in place early resources to steer Kew in that direction.

Professor Sir Ghilliean Prance and his team in the 1990s galvanised the concept of the Millennium Seed Bank Project and secured the funding to build the Millennium Seed Bank at Kew's Wakehurst Place in West Sussex. He handed on the baton in 1999 to Professor Sir Peter Crane just prior to completion of the building and Sir Peter led the Kew team during the critical first half of the Millennium Seed Bank's initial decade of operation, when partnerships were forged around the world in more than 50 countries.

It has been my great pleasure to help the Kew team steer through the second half of the Project's foundation decade, when it achieved its objectives on time and budget. We have also planned the next decade, built on the success of the first, with even more ambitious and broader targets, having proved the value and continued relevance of a global seed bank

partnership for Earth's plant diversity. The Millennium Seed Bank Partnership is now firmly embedded as part of Kew's core work and integrated with other parts of the organisation.

This project has required visionary, energetic and practical leadership, as well as considerable diplomacy and flair in enthusing and inspiring people from many countries. I pay special tribute to the two heads of Kew's Millennium Seed Bank, Dr Roger Smith OBE and Dr Paul Smith, each of whom has displayed such sterling leadership qualities. They, their teams and international partners have much to be proud of in making a real difference to the lives of people and plants on Earth.

At Kew, we have embarked recently upon the Breathing Planet Programme, a step-change in our mission to inspire and deliver science-based plant conservation worldwide, enhancing the quality of life. Banking and sustainably using seed is one of seven core strategies to achieve this mission, alongside scientifically documenting and accelerating access to information on plant diversity, identifying where plants most need our help, working with partner countries to conserve remaining wild vegetation, sustainably using local plants for local people, repairing and restoring plant diversity for human benefit, and inspiring care and sustainable use of plant diversity through our World Heritage botanical gardens and programmes, onsite and through the internet.

Sir David Attenborough has kindly commented that 'Kew's Breathing Planet Programme is perhaps the most exciting, urgent and necessary conservation and sustainability initiative ever.' In planning the second decade of the Millennium Seed Bank Partnership with Paul Smith and his team, I and Kew's Trustees set three primary targets – to bank 25% of the world's plant diversity by 2020, to greatly expand the sustainable use of seed for human benefit, and to place the Partnership on a solid financial footing for its prolonged future. As you will have read in the book, the Millennium Seed Bank was supported during its construction and first decade of operation through a combination of generous grants from the Millennium Commission, the Wellcome Trust, Orange plc, the British Government through Kew's sponsor the Department for Environment, Food and Rural Affairs (Defra), and many other donors listed below.

Securing funding beyond this exciting establishment phase was by no means certain as recently as 2010. However, at the end of that year, we were delighted to receive from the British Government, through Defra, a budget settlement that ring-fenced £3 million for each of the next four years for the Partnership. This grant-in-aid ensures the continuity of staff employment, and care and maintenance of the collection, providing the

foundation from which the team are now able to work with global partners to build a bright future for the Partnership worldwide. Indeed, many partner countries have embraced seed banking beyond Kew's early dreams, building their own seed banks and allocating significant resources, human and financial, to guarantee that we do not let slip through our fingers this most precious gift of living natural heritage.

It remains for me to wholeheartedly congratulate all involved in the production of this fine book. On behalf of the Trustees, staff, students and volunteers of Kew, I commend it to all interested in reading about one of the greatest and most urgent conservation projects underway on Earth, in a time of unprecedented global change.

Professor Stephen D. Hopper FLS
Director (CEO and Chief Scientist)
Royal Botanic Gardens, Kew

We would like to thank the following individuals and organisations for their major contributions to the partnership:

The Millennium Commission
The Wellcome Trust
The Department for Environment,
　Food and Rural Affairs
Orange

Arcadia
The Esmée Fairbairn Foundation
Maite Garcia-Urtiaga
Richard and Kara Gnodde
The Philecology Trust

The American Society for the Royal Botanic Gardens, Kew
The Ann Brown Charitable Settlement
Anglo American
AXA UK
British Airways
Buffini Chao Foundation
The Dennis Curry Charitable Trust
The Dulverton Trust
Natural England (formerly English Nature)
The Eranda Foundation
The Ernest Kleinwort Charitable Trust
GlaxoSmithKline
Goldman Sachs
The HDH Wills 1965 Charitable Trust
Stephen Hester
The John Coates Charitable Trust
The John Ellerman Foundation

The Joseph Banks Society
John Laing
The Kirby Laing Foundation
George and Angela Loudon
The Marisla Foundation (formerly The Homeland Foundation)
Marks & Spencer
Mary Pera
The Negaunee Foundation
The Rufford Foundation
Bryan Sanderson
Sfumato Foundation
Shell International
Robert and Patricia Swannell
David Tate
Tate & Lyle
The Vodafone Foundation

Hands holding *Combretum fragrans* seeds, Burkino Faso.

Index

Bold numbers indicate illustrations

Aberdare Mountains, Kenya **60**
Abrus precatorius **14**
Acacia **14**, **15**, **123**, **131**
Acanthus syriacus 52, **70**, **71**, 131
Access and Benefit Sharing Agreement (ABSA) 58
Acer pseudoplatanus 43
Achillea ptarmica 72
Adansonia digitata 66
Adenium obesum 147
African satinwood 155
Afzelia africana 56
agriculture, in early times 18
Allocasuarina tessellata **14**
almonds, *Amygdalus communis* 17
Aloe ballyi 147
    A. dichotoma 82, **85**
Alstroemeria **87**, 89
Amaryllidaceae 62, 170
Ammocharis **118**, 119
Amyema cambagei 163
Anagallis minima 72
Anemone nemorosa **128**, 129
APG III, plant classification system 151
Ariocarpus kotschoubeyanus **23**
Astragalus physocalyx 57
Aulax palasia **15**
Australia, international partnership 57, 60, 74
Azadirachta indica 66

bamboo seed, African **49**
banana, Yunnan **49**, 57
Banksia brownii, feather-leaved Banksia 162
baobab 66
Baumea acuta **14**
Berchemia idscolor 166
biodiversity hotspot, Sperrgebiet, Namibia 82
black iris 57
Bloomers Valley, UK, habitat restoration **176**
Botswana, international partnership 20, 166, 167
bottle gourd 181
brandybush 155
Brazilian Insitute on Amazon Research, Manaus 43
bromeliads 86
Bulbine crassa, leek lily 57
Bulgaria, international partnership 57, 68
Burkina Faso, international partnership 56
bushman candle 82

cacti 86, 89, 170, **171**
Canada, international partnership 68
Canarium **27**
Cannomois virgata 172
Cardamine gunnii 75
Casuarina cunninghamiana 163
Catharanthus roseus **53**
Centaurea melitensis **14**
chaffweed 72
Chicago Botanic Garden, USA 148, 163
Chile, international partnership 56, 62, **86–89**
    propagation of threatened plants 170, 171

Chilean wine palm 30
China, international partnership 57
    threatened species 78
Chinese Academy of Sciences 57, 145, 164
Chloraea 89
Cleome gynandra 181
climate change 11–13, 132
coco de mer palm **15**, 94
Coffea arabica, coffee bean **15**
colonche, beverage 157
Commiphora seedlings **150**
conservation assessments 73
conservation, *in situ*, *ex situ*, definitions 42
Convention on Biological Diversity (CBD) 22, 31, 35, 53
Coode-Adams, Giles **185**
Crinodendron hookerianum seedlings **150**
cryopreservation techniques 43, 137
cucumber tree 155
cycad 30
Cyrtocarpus procera **157**

daisy, *Bellis perennis* 52
Dalea asurea 56, 170
Damasonium alisma **46**, **47**
Darwin Initiative, Defra 66, 80
Datura quercifolia, oak leaf thorn-apple **76**
dawn redwood 165
Dendrosenecio battiscombei 52
Defra, UK 142
desert rose 147
Difficult Seeds Project 142, 180–181
digitisation of plant information at Kew 151
dormant seeds 125
drift seeds 27
Dutch East India Company 120
dwarf eelgrass 72

Earth Summits, International 22, 35, 58, 80
Echinopsis chiloensis 62, **89**, **170**, **171**
economic uses, plants 152–157, 166
Eden project 148
Egyptian thorn tree 131
Elaeis guineensis 18
Encephalartos altensteinii 30
endemic species, definition 44
endospermic seeds 116
Ensete ventricosum seedlings **150**
Erica 44, 45, **158**
Eriosyce subglobosa **171**
Erlangea remifolia 20, 21
Eucalyptus 163
Euclea divinorum 166
Eucomis autumnalis **157**
Eulychnia castanea **88**
euphorbia 85
European Seed Conservation Network (ENSCONET) 40

Food and Agriculture Organisation (FAO) 142, 180
food crops 19
framework species, restoration ecology 178

fynbos 56, 120, 122, 158

Gentiana 117
Georgia, international partnership 68
germination testing 104
germination, rose family 125
Germinator Predictor database 125
Germplasm Bank of Wild Species, Kunming, China 25, 57, 69, 78, **145**
Gesneriaceae 78
Ginkgo biloba, maidenhair tree 30, 164
Global Crop Diversity Trust 39
global positioning system (GPS), use of 77, 86
Global Strategy for Plant Conservation (GSPC) 22, 31
Graphic Information Systems (GIS) 32, **178**, **179**
Great Basin Native Plant Selection and Increase Project, USA 163
Great Plant Hunt schools project 107
Grewia bicolor, seeds **125**, *G. flava* **155**

habitat restoration 25, 63, 158–163, 176–179
halfmens 82, 85
heathland restoration, Greenham Common, UK 47
Heslop-Harrison, Jack 34, **184**
Hoodia 82, 167
Hooker, Joseph 30
horse chestnut, *Aesculus hippocastanum* 131
horned pondweed 72
Humboldt, Alexander von 30
hunter-gatherer diets 17
Hyphaene palm 77
Hyphaene petersiana 166

Illicium simonsii fruit **78**
Internation Union for the Conservation of Nature (IUCN) 30, 44
Iris nigricans 57
IUCN Species Survival Commission 73
Jan Teerlink's seeds 94, **120–123**
Jordan, international partnership 57, 64
Jubaea chilensis 30, **33**
Judean date palm 42
Juniperus phoenicea 59

kalanchoes 60, 73
karoo habitat 56
Kenya, international partnership 60, 72, 146
Kew Herbarium 31, 151
Kew Innovation Unit 37
Kigelia africana 155
Kirstenbosch National Botanical Gardens 45, 158
kokerboom 82
Krygyzstan, international partnership 68
Kunming Institute of Botany, China 25, 57, 78, 140, 147, 164

Lady Bird Johnson Wildflower Center, USA 148
Lagenaria siceraria 181
Lebanese Agricultural Research Institute 58, 70
Lebanon, Mount **54**, **55**
Leningrad, first seed bank 34, 39
Leucadendron levisanus **127**

Leucospermum 123
Liparia villosa **123**
Lippia chevalieri 156
Lithops ruschiorum 82
Lodoicea maldivica **15**
Lowveld Botanical Garden, South Africa 152
Lupinus pilosus seedlings **150**

Madagascan periwinkle 52
Madagascar, endemic plants 45
   international partnership 66
mahogany, African 56
Malawi, international partnership 56
Mali, Institut D'Economie Rurale 66, 152
mango, *Mangifera indica* **130**
Masson, Francis 30
Medicago **15**
Menodora linoides **170, 171**
Metasequoia glyptostroboides 165
Mexico, Autonomous National University (UNAM) 54
Microcachrys tetragona **65**
Middle-East, habitat restoration 160, 161
mikvetch 57
Millennium Ecosystem Assessment 11
Mini Seed Bank **107**
mistletoe 163
Montserrat, restoration ecology 33
Moringa hildebrandtii, M. oleifera 45
mungongo tree 48
Musa itinerans **49**, 57
Muséum National d'Histoire Naturelle, France 31, 78
Mutisia 86

Namibia, international partnership 65, **82– 85**
National Archives UK, seeds in 121
National Center for Genetic Resources, USA 39
Natural England 46, 186
nawonga 56
neem 66
Nematolepis wilsonii 59

oak, English, *Quercus robur* 130, **131**
oil palm 18
Oldfieldia dactylophylla 56
olive, *Olea europaea* 150, **151**
Orchid Seed Stores for Sustainable Use (OSSSU) 80
orchids, *Chloraea* 89
orchids, threatened species 80, 81
orthodox seeds 42, 130
Osyris lanceolata 55, 179
Overcus, acorns 17
Oxytenanthera abyssinica, seed **49**

Pachypodium namaquanum 82, 85
palm, vegetable ivory or ilala 77
Panax pseudoginseng 165
Paphiopedilum rothschildianum 80
Paraisometrum mileense 78, 162
pea, creeping darling, Sturt's desert 74
Phytophthora fungal disease 162
pigeon wood 157
pineapple flower, pineapple lily 157
Placea lutea 62, 170
plant collectors, historic, at Kew 30
Plant Conservation Strategies, International Diploma 147
Plant Conservation Techniques Course at Kew 147
Plant Germplasm Conservation Research Group, Natal 43
plant-conservation targets 58
plants, economic uses 152–157, 166

Plazia cheiranthifolia **63**
Prance, Sir Ghillean 30, **184**
priming of seeds 134, 137
Primula 117
Protea cryophila, snow protea 56
Psorospermum febrifugum 156
Puya 86, *P. chilensis* **88**

quiver tree 82, 85

recalcitrant seeds 42, 130
Red List of Threatened Species, IUCN 30, 44, 82, 147
Regent honeyeater 25, 163
restoration ecology 25, 33, 178 *see also habitat restoration*
Rhododendron 117
river oak 163
rose family, germination 125
Royal Botanic Gardens, Melbourne, Australia 59
Rubiaceae 32

saltbush, *Salsola nollothensis* 65, 85
Sarcocaulon marlothii 82
sausage tree 155
Save our Seeds project 107
Schinziophyton 4
Schizanthus **88**
Schoenoplectus triqueter 57, 186, **187**
sea beans 27
seed
   accessions 39
   biochemistry 134, 135
   collection guides 140, 151
   gathering and identification overview 76
   morphology 26, 27
seed banking technology 110–113
seed collecting 52–55
   types of partner organisation 72
Seed Conservation Techniques Course, Kew 140, 148
seed conservation technology 126
seed germination 172
Seed Information Database, MSBP 26, 150, 172
seed oils 137
seed processing 100–105
seed research 93, 114, 115, 132
seed vault, Wellcome Trust Millennium Building **93**, 95
seed viability, testing 102
seed viability and glutathione 134
seed-exchange programme 34
seeds, effects of drying 130, 131, in early Neolithic sites 39,
   largest in MSBP 77, longevity 116, 117, 134, 135, orthodox
   42, recalcitrant 42, smallest 81, under stress 137
Seeds for Life project, Kenya 146, 148
Seeds of Success, USA 62, 163
Senecio 52, **60**, 86
Sesame, *Sesamum abbreviatum* 85
Shanghai Expo, British Pavilion **165, 174, 175**
shea butter tree 66, 130, 137, 181
slipper orchids 80
Slovakia, international partnership 68
Smith, Paul 10, **11, 13**, 46, 47
Smith, Roger 41, **185**
sneezewort 72
sourplum, large 155
South Africa, international partnership 56
South African National Biodiversity Institute (SANBI) 140, 158
South Australian Centre for Rare and Endangered (SACRED)
   Seeds Project 74
spade-leaf bittercress 75
Species Targeting Team, MSBP 73

Spruce, Richard 30
starfruit **46**, 47
Stenocereus stellatus **157**
Sterculia setigera seeds **182**
Stipagrostis giessii **15**
Stomatostemma monteiroae 166
Strelitzia victoria-reginae seeds **15**
Survey of Economic Plants for Arid and Semi-Arid Lands
   (SEPASAL) 45
Svalbard Global Seed Vault 39
Swainsona formosa, S. viridis 74
sycamore 43, 130
Syrian bear's-breech 52, **70, 71**
tallgrass prairie, USA, habitat restoration 63

Tanzania, international partnership 60
Tasmania, Seed Safe project 57
Technology and Training Team, MSBP 142
Terminalia seeds **41**
Thatcher, Margaret **184**
Thompson, Peter 40, **184**
threatened Chilean flora 170, 171
threatened species, in China 78, 79, orchids 80
Threatened Species Classification, IUCN 44
Tibilisi Botanical Garden and Institute of Botany, Georgia
   58, 68
traditional Chinese medicine (TCM) 165
traditional medicines 152, 155
training programmes, MSBP and Kew 140–149
Trema orientalis 157
triangular club rush 57, 186, **187**
Tsodilo daisy 20, 21

UK Flora Programme at Kew 40
UK Overseas Territories (UKOT) 32
University of Kwazulu, Natal 43
USA, international partnership 62, 63, 163
Useful Plants Project 62, 152–155

Vavilov Institute of Plant Industry, Russia 39
Vavilov, Nikolai 34, 39
Velleia rosea **15**
Velnos's Vegetable Syrup 74
velvet raisin 155
Veronia 36
Vitellaria paradoxa 66, 130, **137**, 181
voucher, plant **32**, 76

Wakehurst Place **98–99**
Wellcome Trust Millennium Building **92–99**
Welwitschia mirabilis **50, 51**, 82
Western Australian Department of Environment and
   Conservation 148, 162
white box 163
white rampion 187
Widdringtonia schwarzii **168**, 169
Willowmore cedar **168**, 169
Wilson, Ernest 30
Wollemi pine 30
wood anemone **128**, 129
Worldclim database 126

Xanthomyza phrygia 25, 163
Ximenia caffra 155
xoconostle 157

Zannichellia palustris 72
Zanthoxylum gilletii 155
Zostera noltii 72

Acknowledgements

The authors would like to thank everyone at the Millennium Seed Bank Partnership and the Royal Botanic Gardens, Kew, who gave their time to be interviewed and provided valuable information and advice during the writing of this book.

The publishers would like to thank the many MSBP staff and staff members in partnership organisations who have generously provided photographs to be reproduced in this book. Without their help this book would not have been possible and we regret that space does not permit the listing of all individual photography contributors. Special thanks are extended to Andrew McRobb, Kew's photographer, who provided many of the images in this book and undertook valuable picture research.

Thanks also to Paul Smith of the MSBP, who provided much appreciated advice and support for the duration of the project. We are grateful to Wolfgang Stuppy of the MSBP for permitting many of his photographs to be used throughout the book and for sourcing further images, as well as for writing caption text together with his colleague Michiel van Slageren. Particular thanks also go to Michiel for his detailed reading of the proofs.

Photography credits

While every effort has been made to trace and acknowledge copyright holders, we would like to apologise in advance for any errors or omissions.

16 top right © Ray Witlin/World Bank

20 left © 2011 photolibrary.com

20 centre bottom, centre right © 2011 photolibrary.com

21 top left © 2011 shutterstock

26 © Christina Murrey

41 left © 2011 photolibrary.com

41 right © Mari Tefre

47 bottom © 2011 photolibrary.com

63 © Jonathan Drori

69 © John Millar

87–89 all images © Charles Godfray

120 map © David Rumsey map collection

120 © iStockphoto.com/Keith Reicher

122 top centre © Jeff Eden

149 portrait of Michael Eason © Christina Murrey.